Mechatronics and manufacturing engineering

WOODHEAD PUBLISHING REVIEWS:

MECHANICAL ENGINEERING

Series Editor: Professor J. Paulo Davim,
Department of Mechanical Engineering, University of Aveiro,
Portugal and Head of MACTRIB – Machining and Tribology Research Group
(email: pdavim@ua.pt)

Woodhead Publishing is pleased to publish this major Series of books entitled *Woodhead Publishing Reviews: Mechanical Engineering*. The Series Editor is Professor J. Paulo Davim, Department of Mechanical Engineering, University of Aveiro, Portugal and Head of MACTRIB – Machining and Tribology Research Group.

This research Series publishes refereed, high quality articles with a special emphasis on research and development in mechanical engineering from a number of perspectives including (but not limited to): machining and machine tools; tribology and surface engineering; materials and manufacturing processes; solid mechanics and structural mechanics; computational mechanics and optimization; mechatronics and robotics; fluid mechanics and heat transfer; renewable energies; biomechanics; micro- and nano-mechanics, etc. We seek authors, editors and contributors from a broad range of areas within the mechanical engineering discipline.

This Series examines current practises and possible future developments within the research field and industry-at-large. It is aimed at an international market of academics, practitioners and professionals working in the area. The books have been specially commissioned from leading authors, with the objective of providing the reader with an authoritative view of current thinking.

New authors: we would be delighted to hear from you if you have an idea for a book. We are interested in both shorter, practically orientated publications (45,000+ words) and longer, theoretical monographs (75,000–100,000 words). Our books can be single, joint or multi-author volumes. If you have an idea for a book, please contact the publishers or Professor J. Paulo Davim, the Series Editor.

Dr Glyn Jones
Woodhead Publishing Limited
Email: gjones@chandospublishing.com
www.woodheadpublishing.com

Professor J. Paulo Davim
Department of Mechanical Engineering, University of Aveiro, Portugal
Email: pdavim@ua.pt
http://www2.mec.ua.pt/machining/pers-davim.htm

Woodhead Publishing Limited: established in 1989, Woodhead Publishing is a leading independent international publisher, publishing in the following main areas: food science, technology and nutrition; materials engineering; welding and metallurgy; textile technology; environmental technology; finance, commodities and investment; and mathematics.

Our ambitious publishing plans for the future will continue to bring you a range of authoritative reference books, professional texts and monographs, all written and produced to the exacting standards that have made Woodhead Publishing one of the UK's fastest growing independent publishers. All of our books are written in direct response to customers' needs by a truly international team of authors, ensuring they are designed for and relevant to a global audience.

Woodhead Publishing books are available worldwide, either direct, via our website, www.woodheadpublishing.com, or through booksellers and an international network of agents and representatives.

Professor J. Paulo Davim received his PhD in Mechanical Engineering from the University of Porto in 1997 and the Aggregation from the University of Coimbra in 2005. Currently, he is Aggregate Professor in the Department of Mechanical Engineering of the University of Aveiro and Head of MACTRIB - Machining and Tribology Research Group. He has more than 25 years of teaching and research experience in manufacturing, materials and mechanical engineering with special emphasis in machining and tribology. He is the Editor of four international journals, and also guest editor, editorial board member, reviewer and scientific advisor for many international journals and conferences. He has also published, as author and co-author, more than 30 book chapters and 300 articles in ISI journals (h-index 17) and conferences.

Bulk orders: some organisations buy a number of copies of our books. If you are interested in doing this, we would be pleased to discuss a discount. Please email wp@woodheadpublishing.com or telephone +44 (0) 1223 499140.

Related titles:

Materials and surface engineering: Research and development
Number 2 in the Woodhead Publishing Reviews: Mechanical Engineering Series
(ISBN 978-0-85709-151-2)
This book, the second in the **Woodhead Publishing Reviews: Mechanical Engineering Series**, will present full research articles, reviews and cases studies with a special emphasis on the research and development of materials, and surface engineering and its applications. Surface engineering techniques are being used in the automotive, aircraft, aerospace, missile, electronic, biomedical, textile, petrochemical, chemical, moulds and dies, machine tools, and construction industries. Materials science is an interdisciplinary field involving the micro and nano-structure, processing, properties of materials and its applications to various areas of engineering, technology and industry. All types of materials are addressed including metals and alloys, polymers, ceramics and glasses, composites, nano-materials, biomaterials, etc. The relationship between micro and nano-structure, processing, and properties of materials will be discussed. Surface engineering is a truly interdisciplinary topic in materials science that deals with the surface of solid matter.

Machining and machine-tools: Research and development
Number 3 in the Woodhead Publishing Reviews: Mechanical Engineering Series
(ISBN 978-0-85709-154-3)
The third book in the **Woodhead Publishing Reviews: Mechanical Engineering Series** includes high quality papers with a special emphasis on research and development in machining and machine-tools. Machining and machine tools is an important subject with application in several industries. Parts manufactured by other processes often require further operations before the product is ready for application. Traditional machining is the broad term used to describe removal of material from a workpiece, and covers chip formation operations including: turning, milling, drilling and grinding. Recently the industrial utilization of non traditional machining processes such as EDM (electrical discharge machining), LBM (laser-beam machining), AWJM (abrasive water jet machining) and USM (ultrasonic machining) has increased. The performance characteristics of machine tools and the significant development of existing and new processes, and machines, are considered. Nowadays, in Europe, USA, Japan and countries with emergent economies machine tools is a sector with great technological evolution.

Details of these and other Woodhead Publishing books can be obtained by:

- visiting our web site at www.woodheadpublishing.com
- contacting Customer Services (e-mail: sales@woodheadpublishing.com; fax: +44(0) 1223 832819; tel: +44(0) 1223 499140; address: Woodhead Publishing Limited, 80 High Street, Sawston, Cambridge CB22 3HJ, UK)

If you would like to receive information on forthcoming titles, please send your address details to Customer Services, at the address above. Please confirm which subject areas you are interested in.

Mechatronics and manufacturing engineering

Research and development

EDITED BY
J. PAULO DAVIM

WOODHEAD PUBLISHING

Oxford Cambridge Philadelphia New Delhi

Published by Woodhead Publishing Limited, 80 High Street, Sawston, Cambridge CB22 3HJ, UK
www.woodheadpublishing.com

Woodhead Publishing, 1518 Walnut Street, Suite 1100, Philadelphia, PA 19102–3406, USA

Woodhead Publishing India Private Limited, G-2, Vardaan House, 7/28 Ansari Road, Daryaganj, New Delhi – 110002, India
www.woodheadpublishingindia.com

First published 2012, Woodhead Publishing Limited

British Library Cataloguing in Publication Data
A catalogue record for this book is available from the British Library.

Library of Congress Control Number 2011942127

Woodhead Publishing ISBN 978-0-08-101654-1 (print)
ISBN 978-0-85709-589-3 (online)

ISSN 2048-0571 Woodhead Publishing Reviews: Mechanical engineering (print)
ISSN 2048-058X Woodhead Publishing Reviews: Mechanical engineering (online)

Typeset by RefineCatch Limited, Bungay, Suffolk

Contents

List of figures

List of tables

Preface

Nowadays, it is current to define mechatronics as an interdisciplinary area of engineering that combines mechanical, electronic and computer engineering. The main objective of this engineering field is the study of automata from an engineering perspective, thinking on the design of products and manufacturing processes. Today, mechatronics systems are well established in a great number of industries such as automotive, aircraft, aerospace, machine tools, product manufacturing, computers, electronics, semiconductor and communications, biomedical, etc.

This book aims to provide information on mechatronics for modern industry with an emphasis on manufacturing engineering. Chapter 1 looks at issues of innovation, design and development in relation to the implementation of light-scattering instrumentation. Chapter 2 is dedicated to recent developments for planar micromanipulation on microconveyor platforms. Chapter 3 presents basic active/passive joint torque control of single-axis arm designed with an ultrasonic motor. Chapter 4 covers signal processing for tool condition monitoring: from wavelet analysis to sparse decomposition. Chapter 5 is dedicated to ANN modelling of fractal dimension in machining. Chapter 6 contains information on soft computing techniques for predicting forces and damage in drilling of polymer composites. Chapter 7 covers integrating surface response methodology with particle swarm optimisation for minimising burr size in drilling. Finally, Chapter 8 is dedicated to single point incremental forming of polymers.

The book can be used as a research tool for final undergraduate engineering courses or as a topic on mechatronics at the postgraduate level. It can also serve as a useful reference for academics, mechatronics and manufacturing researchers, mechanical, mechatronics, manufacturing, electronic and computer engineers, and professionals in mechatronics and related industries. The scientific interest in this book is evident for many important centres of research, laboratories and universities throughout the world. Therefore, it is hoped this book will inspire and enthuse other research in this field.

The Editor acknowledges Woodhead Publishing for this opportunity and for their enthusiastic and professional support. Finally, I would like to thank all the chapter authors for their availability for this work.

J. Paulo Davim
University of Aveiro, Portugal
May 2011

About the contributors

J. Paulo Davim (Editor)

J. Paulo Davim received his PhD in mechanical engineering from the University of Porto in 1997 and the Aggregation from the University of Coimbra in 2005. He is currently Aggregate professor in the Department of Mechanical Engineering of the University of Aveiro and Head of MACTRIB – Machining and Tribology Research Group. He has more than 25 years of teaching and research experience in manufacturing, materials and mechanical engineering, with special emphasis in machining and tribology. He is the editor of nine international journals, guest editor, editorial board member, reviewer and scientific adviser for many international journals and conferences. He has also published more than 30 book chapters and 300 articles as author and co-author in refereed international journals and conferences.

J. Paulo Davim
Department of Mechanical Engineering, University of Aveiro
Campus Santiago, 3810-193 AVEIRO
PORTUGAL
e-mail: pdavim@ua.pt

Chapter 1

Roger White

Roger White graduated from Bath University with a PhD in materials science for research on creep failure mechanisms in polymer composites before becoming a group leader for THORN-EMI CRL in Optical Sensor R&D. He then moved into technical marketing in the analytical instrumentation industry supporting macromolecular separations system sales, managing detector production and developing laboratory automation. He then started a polymer analysis business marketing imported state-of-the-art systems, undertaking commercial contract analytical research and participating in R&D collaborations to develop new methods. He was a member of the DTI Faraday Foresight Lab-on-a-Chip project and is a past chairman of the Welsh Optoelectronics Forum, where he helped develop the OpTIC Technium.

Roger White
Norton Scientific Inc.
3430 Schmon Parkway
Thorold, Ontario L2V 4Y6
CANADA
e-mail: rjw@norsci.ca

David Bradley

David Bradley has been involved in the design, development and implementation of mechatronic devices and systems since the mid-1980s and has included work in mechatronic design methods and methodology, robotic technologies for construction and physiotherapy, system modelling and instrumentation systems. Current work includes the design of intelligent and mechatronic systems, system modelling, automated systems for physiotherapy and applications in telecare and telehealth. He has been an invited speaker at various international conferences and workshops, most recently in China, South Africa and Colombia. He is a professorial consultant at the University of Abertay, Dundee, a visiting professor at Sheffield University, a fellow of the IE7 and a founder member of the Mechatronics Forum.

David Bradley (Corresponding author)
CES
University of Abertay Dundee
Bell Street
Dundee DD1 1HG
UNITED KINGDOM
e-mail: d.bradley@abertay.ac.uk

Nikolai Zhelev

Nikolai Zhelev is an honorary professor in eight universities in the UK, China and Bulgaria. He is a fellow of the Institute of Biology in the UK, a full member of the Bulgarian National Academy of Medicine, and *doctor honoris causa* at the Medical University Plovdiv. Professor Zhelev has been involved in founding four start-up biotech companies and has worked as head of the departments of biochemistry, cell biology and proteomics in Cyclacel Pharmaceuticals Inc. He is a member of the editorial boards of four scientific journals and has published his work on cancer drug development in a number of high impact journals including *Nature Medicine* and *Nature Biotechnology*. Professor Zhelev's current research includes systems biology in cancer.

Nikolai Zhelev
SIMBIOS
University of Abertay Dundee
Bell Street
Dundee DD1 1HG
UNITED KINGDOM
e-mail: n.zhelev@abertay.ac.uk

Chapter 2

Panos Lazarou

Panos Lazarou is a member of the Robotics Group of the Mechanical and Aeronautics Engineering Department, University of Patras, Greece. He received his BSc degree in physics in 2002 and his PhD from the Department of Mechanical Engineering and Aeronautics in 2010, both from the University of Patras. His main research interests include microassembly, micromanipulation and micromechatronics. His current research is in the field of electrostatic self-assembly and programmable force fields for micromanipulation.

Panos Lazarou (Corresponding author)
Robotics Group, Department of Mechanical Engineering & Aeronautics
University of Patras
TK 36500
GREECE
e-mail: plazarou@freemail.gr

Nikos A. Aspragathos

Nikos A. Aspragathos leads the Robotics Group of the Mechanical and Aeronautics Engineering Department, University of Patras, Greece. He received his electrical engineering diploma (1975) and PhD (1981) both from University of Patras. He was a visiting member of the staff in UWIST, Cardiff, from 1987–88. His main research interests are robotics, intelligent motion planning and control for dextrous manipulation, knowledge-based design, microsystems assembly automation and computer graphics. He is a reviewer in a considerable number of journals and member of the editorial boards of *Mechatronics Journal, Vimation Journal* and the *International Journal of Automation and Control.* He has published more than 150 papers in journals and conference proceedings. He has been and is currently involved in research projects funded by Greek and European Union sources.

Nikos A. Aspragathos
Robotics Group, Department of Mechanical Engineering & Aeronautics
University of Patras
TK 36500
GREECE
e-mail: asprag@mech.upatras.gr

Chapter 3

Fusaomi Nagata

Fusaomi Nagata received his BEng degree from the Department of Electronic Engineering at Kyushu Institute of Technology in 1985, and his DEng degree from the Faculty of Engineering

Systems and Technology at Saga University in 1999. He was a research engineer with Kyushu Matsushita Electric Co. from 1985–88, and a special researcher with Fukuoka Industrial Technology Centre from 1988–2006. He is currently an associate professor at the Department of Mechanical Engineering, Faculty of Engineering, Tokyo University of Science, Yamaguchi, Japan. His research interests include intelligent control of industrial robots and its applications.

Fusaomi Nagata (Corresponding author)
Department of Mechanical Engineering
Tokyo University of Science
Yamaguchi
1-1-1 Daigaku-Dori
Sanyo-Onoda 756-0884
JAPAN
e-mail: nagata@ed.yama.tus.ac.jp

Keisuke Ogiwara

Keisuke Ogiwara received his BEng degree from the Department of Electronic Engineering at Tokyo University of Science, Yamaguchi in 2009. He is currently a master's degree student at the Faculty of Engineering, Tokyo University of Science, Yamaguchi. His research interest is assistance devices for physically handicapped persons.

Keisuke Ogiwara
Department of Mechanical Engineering
Tokyo University of Science
Yamaguchi
1-1-1 Daigaku-Dori
Sanyo-Onoda 756-0884

JAPAN
e-mail: ogiwara821@yahoo.co.jp

Keigo Watanabe

Keigo Watanabe received his BEng and MEng in mechanical engineering from the University of Tokushima, Tokushima, Japan, in 1976 and 1978, respectively. He received his DEng in aeronautical engineering from Kyushu University, Fukuoka, Japan, in 1984. From April 1998 to March 2009, he was with the Department of Advanced Systems Control Engineering, Saga University. He is now with the Department of Intelligent Mechanical Systems, Graduate School of Natural Science and Technology, Okayama University. He has published more than 600 technical papers in transactions, journals and international conference proceedings, and is the author or editor of 25 books.

Keigo Watanabe
Department of Intelligent Mechanical Systems
Graduate School of Natural Science and Technology
Okayama University
3-1-1 Tsushima-naka, Kita-ku
Okayama 700-8530
JAPAN
e-mail: watanabe@sys.okayama-u.ac.jp

Chapter 4

Zhu Kunpeng

Zhu Kunpeng received his PhD degree in 2007 from the Department of Mechanical Engineering, National University of Singapore (NUS). He is currently a research fellow in the Department of Mechanical Engineering, NUS. His research interests are in the areas of precision machining modelling, tool condition monitoring, signal processing and engineering informatics. In these areas, he has published in refereed journals and contributed to books. He has been on the editorial board and associate editor of international journals and IEEE conferences.

Zhu Kunpeng (Corresponding author)
Department of Mechanical Engineering
National University of Singapore
Kent Ridge Crescent
11926 SINGAPORE
e-mail: kunpengz@hotmail.com

Wong Yoke San

Wong Yoke San received his PhD degree from the University of Manchester Institute of Science and Technology, Manchester, UK. He is currently a professor with the Department of Mechanical Engineering, National University of Singapore. His research interests are in modelling, simulation, instrumentation and control primarily for applications in manufacturing automation. In these areas, he has published in refereed journals and proceedings of international conferences, contributed to books, and participated in several local and overseas funded projects. He has been the associate editor of various international journals such as *IEEE Trans. Automation Science and Engineering* and the *International Journal of Advanced Manufacturing*.

Wong Yoke San
Department of Mechanical Engineering
National University of Singapore
Kent Ridge Crescent
11926 SINGAPORE
e-mail: mpewys@nus.edu.sg

Hong Geok Soon

Hong Geok Soon is currently an associate professor in the Department of Mechanical Engineering at the National University of Singapore. He received his PhD degree in control engineering in 1987 from the University of Sheffield, UK. The topic of his research was in modelling and control of mechanical systems, tool condition monitoring, and AI techniques in monitoring and diagnostics. He has published extensively in refereed journals and contributed to books in these areas.

Hong Geok Soon
Department of Mechanical Engineering
National University of Singapore
Kent Ridge Crescent
11926 SINGAPORE
e-mail: mpehgs@nus.edu.sg

Chapter 5

Prasanta Sahoo

Prasanta Sahoo is a professor in the Department of Mechanical Engineering, Jadavpur University, Kolkata,

India. His main research interests include tribology of materials and structural mechanics. He has authored a textbook on engineering tribology and a number of book chapters. He has co-authored more than 200 technical papers. He is the associate editor of two international journals, member of the editorial boards of five international journals and on the review boards of more than 30 international journals.

Prasanta Sahoo (Corresponding author)
Department of Mechanical Engineering
Jadavpur University
Kolkata 700032
INDIA
e-mail: psjume@gmail.com

Tapan Kr. Barman

Tapan Kr. Barman is an assistant professor in the Department of Mechanical Engineering, Jadavpur University, Kolkata, India. His main research interests include roughness characterisation and ANN modelling. He has co-authored more than 40 technical papers in international journals and conference proceedings.

Tapan Kr. Barman
Department of Mechanical Engineering
Jadavpur University
Kolkata 700032
INDIA

Chapter 6

Inderdeep Singh

Inderdeep Singh is assistant professor with the Department of Mechanical and Industrial Engineering, Indian Institute of Technology, Roorkee, India. He received his PhD degree from the Indian Institute of Technology (IIT), Delhi. His research work won an award from the Foundation for Innovation and Technology Transfer (FITT) as the 'Best Industry Relevant PhD Thesis of the Academic Year 2004–2005' at IIT Delhi. He is involved in research and development work in the field of machining of composite materials. He has guided one PhD thesis and is currently supervising five PhD scholars. He has also supervised 15 postgraduate students towards successful completion of their dissertation work. He has to his credit 75 research publications in international and national journals and conference proceedings.

Inderdeep Singh (Corresponding author)
Department of Mechanical and Industrial Engineering
Indian Institute of Technology, Roorkee
Roorkee 247667
INDIA
e-mail: inderfme@iitr.ernet.in

Pawan Kumar Rakesh

Pawan Kumar Rakesh graduated in mechanical engineering from Sant Longowal Institute of Engineering and Technology (SLIET), Longowal, Punjab. He did his MTech in manufacturing technology at the Department of Industrial Engineering, National Institute of Technology, Jalandhar. He is currently working as research scholar at the Department of Mechanical and Industrial Engineering, Indian Institute of Technology, Roorkee. He is involved in research and development in the field of primary and secondary processing of polymeric matrix composite materials. He has to his credit 21 research publications in international and national journals and conference proceedings.

P.K. Rakesh
Department of Mechanical and Industrial Engineering
Indian Institute of Technology, Roorkee
Roorkee 247667
INDIA

Jagannath Malik

Jagannath Malik was born in Cuttack (Odisha), India. Currently he is pursuing an integrated dual degree (bachelor's/master's) in electronics and computer engineering with specialisation in wireless communication at the Indian Institute of Technology (IIT), Roorkee, India. His research interests include soft computing techniques, artificial neural networks (ANNs), optimisation algorithms, image processing, machine learning, millimetre-wave engineering, metamaterial, microstrip antennas for communications, RF and microwave designs. He is also a member of the Biomedical Signal Processing Lab at the Department of Electrical Engineering, IIT Roorkee. He has published various papers in the fields of microstrip antenna and predictive modelling of drilling in composites.

Jagannath Malik
Department of Electronics and Computer Engineering

Indian Institute of Technology, Roorkee
Roorkee 247667
INDIA

Chapter 7

V.N. Gaitonde

V.N. Gaitonde is a professor in the Industrial and Production Engineering Department at the BVB College of Engineering and Technology, Hubli. He obtained an ME degree in production management from Karnataka University, Dharwad, and a PhD degree from Kuvempu University, Shimoga. His fields of interest include process modelling and optimisation, application of artificial neural networks (ANN), genetic algorithm (GA), particle swarm optimisation (PSO) and robust design in manufacturing processes. He has more than 22 years of teaching and research experience. He is an editorial board member of four international journals, reviewer for many international journals and has published more than 50 papers in refereed international journals and conference proceedings.

V.N. Gaitonde (Corresponding author)
Department of Industrial and Production Engineering
BVB College of Engineering and Technology
Hubli-580 031
Karnataka
INDIA
e-mail: gaitondevn@yahoo.co.in

S.R. Karnik

S.R. Karnik is an assistant professor in electrical and electronics engineering at the BVB College of Engineering and Technology, Hubli. He obtained an M.Tech. from IIT Kharagpur. His fields of interest include process modelling and optimisation, power system analysis, artificial neural networks (ANN), genetic algorithm (GA), particle swarm optimisation (PSO) and robust design applications to power system monitoring and control. He has more than 24 years of teaching and research experience. He is an editorial board member of four international journals, reviewer for many international journals and has published more than 40 papers in refereed international journals and conference proceedings.

S.R. Karnik
Department of Electrical and Electronics Engineering
BVB College of Engineering and Technology
Hubli-580 031
Karnataka
INDIA
e-mail: karniksr@yahoo.com

J. Paulo Davim

See entry on page xi.

Chapter 8

Maria Beatriz Silva

Maria Beatriz Silva received her PhD in mechanical engineering from the Instituto Superior Técnico, TULisbon,

Portugal, in 2008. She is assistant professor and head of the Material Testing Laboratory of the Manufacturing Unit of the Instituto Superior Técnico. Her research interests include numerical and experimental simulation of metal forming processes and she is co-author of 30 papers in international journals and conference proceedings. In 2008 she was awarded the Donald Julius Groen Prize by the Structural Technology and Materials Group of the Institution of Mechanical Engineers (UK).

Maria Beatriz Silva
Instituto Superior Técnico, TULisbon
Av. Rovisco Pais
1049-001 Lisboa
PORTUGAL

Tânia Marques

Tânia Marques received her MSc in materials engineering from Instituto Superior Técnico, TULisbon, Portugal, in 2010. She is research assistant at the Manufacturing Unit of the Instituto Superior Técnico and her research interests are mainly focused on sheet metal forming.

Tânia Marques
Instituto Superior Técnico, TULisbon
Av. Rovisco Pais
1049-001 Lisboa
PORTUGAL

Paulo A.F. Martins

Paulo A.F. Martins received his PhD in mechanical engineering from Instituto Superior Técnico, TULisbon, Portugal, in 1991, and received his Habilitation in 1999 in recognition of his work in the numerical and experimental simulation of metal forming processes. He is currently professor of manufacturing at Instituto Superior Técnico, TULisbon. His research interests include metal forming and metal cutting and he is co-author of three books, several national and international patents and 200 papers in international journals and conference proceedings. He is on the editorial board and collaborates as reviewer of several international journals.

Paulo A.F. Martins (Corresponding author)
Instituto Superior Técnico, TULisbon
Av.Rovisco Pais
1049-001 Lisboa
PORTUGAL
e-mail: pmartins@ist.utl.pt

1

Implementation of light-scattering instrumentation: innovation, design and development

Roger White, Nikolai Zhelev and David Bradley

Abstract: Product design invokes a complex process which integrates technical issues with user needs, finance and marketing within a product development strategy for the short, medium and long terms. Failure to understand these issues and their integration will significantly impact on the success of the resulting product or system. This implies that non-technical issues, which can sometimes be afforded a lower status, are given an equal weighting from the initiation of the design process.

The chapter looks at this process in the context of a novel form of light-scattering instrument targeted at aggregation analysis. It considers the nature of the application and the need for an instrument of the type proposed along with generic considerations associated with the progression from idea through concept to product and beyond. It then uses the development profile for an advanced protein aggregation monitor, the *Norton NS4910 PAM*, as an illustrative example.

Key words: innovation; design; user requirements; light-scattering; bio-pharmaceuticals; funding strategies.

Glossary

A_2 The second osmotic virial coefficient. The change in the chemical potential on mixing is made up of an ideal term and an excess term such that:

$$\Delta\mu_1 = \Delta\mu_1^{\text{Ideal}} + \Delta\mu_1^{\text{Excess}}$$

The second virial coefficient is then proportional to the excess chemical potential of mixing when:

$$A_2 = \frac{-\Delta\mu_1^{\text{Excess}}}{V_s c^2}$$

c The concentration of the solution under investigation using light-scattering.

$^{d_n}/_{d_c}$ The specific refractive index increment expressed as the change in refractive index with concentration.

mole The mole (mol) is the SI unit of measurement used to express an amount of a chemical substance.

M_w The weight average molar mass describes the relationship between the number of moles and the molar mass.

N_A Avogadro's Number expresses the number of elementary entities per mole and has the value $6.02214179 \times 10^{23}$ mol^{-1}.

n_0 The refractive index of the solvent used expressed as the ratio of the speed of light in a vacuum to that in the solvent.

q The scattering wave vector is a means of describing a wave. Its magnitude is inversely proportional to the wavelength and its direction is normally, but not always, the direction of propagation of the wave.

R_g The root mean radius of gyration calculated as the root mean square distance of the elements making up the relative to a chosen reference such as an axis or the centre of gravity.

$R(\theta)$ The excess Rayleigh scattering factor defined as the Rayleigh ratio of the solution or single particle event from which is subtracted the Rayleigh ratio of the carrier fluid along with any other background contributions.

v_s The partial specific volume of the solvent. This is the inverse of density and is usually expressed in millilitres per gram (mL/g).

λ The wavelength of the laser

1.1 Introduction

The translation of a concept for an advanced instrumentation system into a marketable device involves a complex innovation, design and development process which integrates the underlying technical issues associated with solving the fundamental measurement issues with a requirement to properly and fully understand and integrate user needs, finance and marketing within the product development strategy for the short, medium and long terms. Indeed, it can be argued that a failure to understand these, and related, issues is more likely to impact on the success, or otherwise, of the resulting product or system than the nature of the technology around which they are structured. In practice, this requires that what are often considered as non-technical issues, and which at times are thus afforded a lower status, are in fact given an equal weighting with the technical issues from the initiation of the design process.

The chapter therefore considers the innovation, design and development process as associated with the implementation and introduction of a novel form of light-scattering instrument intended for the bio-pharmaceutical industry and targeted at aggregation analysis. In doing so, the chapter considers the nature of the application and establishes the need for an instrument of the type proposed from a consideration of user need and defines the underlying physical principles on which the instrument itself is based. It then sets out the generic considerations associated with the progression from idea through concept to product, placing these in the context of factors such as innovation strategy, the management of complexity, design methods, requirements capture and analysis, funding, prototype development, marketing and continuous development – including the need to keep ahead of the competition.

In doing so, it builds on experience gained both as product developer and end-user, or client, in the pharmaceutical and process industry sectors along with work on design theory and practice with particular reference to mechatronics.

1.2 Application and need

1.2.1 Fundamentals of static light-scattering

Static light-scattering (Einstein, 1910) is aimed at the characterisation of both synthetic and natural polymers with

Figure 1.1 The basis structure of a light-scattering system

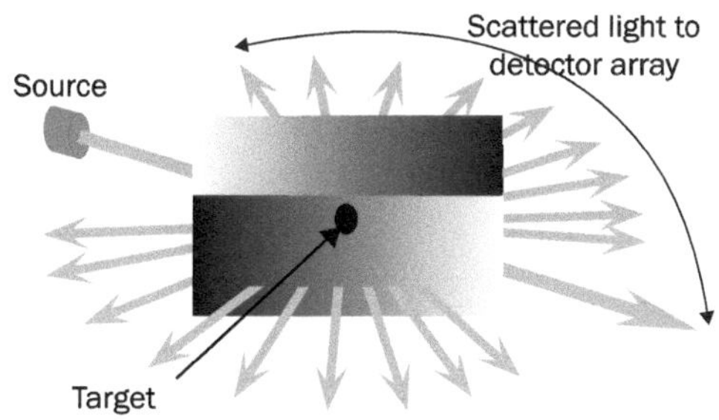

specific reference to determining molar mass distributions. The basic principles are illustrated in Figure 1.1 in which the intensity of the scattered light from the monochromatic source at a particular angle of scattering may be directly related to the weight average molar mass (M_w) and the root mean radius of gyration (R_g) of the molecule or particle population in the cell. These relationships are most commonly defined by the Zimm equation in which $R(\theta,c)$ describes the angular dependence of the scattering process in terms of the system parameters and $P(\theta)$ is a shape factor that accounts for variations in molecular conformation. Calibration is based around the use of a known medium such as toluene for which the Rayleigh Ratio is well defined.

$$\frac{K^{*}c}{R(\theta,c)} = \frac{1}{\left[M_w P(\theta)\right]} + 2cA_2 = \frac{1}{M_w}\left(1+\frac{q^2R_g^2}{3}\right) + 2cA_2 \quad (1.1)$$

where:

$$K^{*} = \frac{4\pi^2 n_0^2 \left({}^{\mathrm{d_n}}\!/\!{}_{\mathrm{d_c}} \right)^2}{\lambda_0^4 \mathrm{N_A}}$$

$$P(\theta) = 1 \Big/ \left(1 + \frac{q^2 R_g^2}{3}\right)$$

and

$$q = 4\pi n_0 \frac{\sin\left(\theta/2\right)}{\lambda_0}$$

In its most simple form for measurements made at low angle and infinite dilution, the Zimm equation reduces to:

$$\frac{K_c}{\Delta R(\theta \to 0,\ c \to 0)} = \frac{1}{M_w} \qquad (1.2)$$

since $P(0) = 1$.

A Zimm plot (Ahmad et al. 2009; Zimm, 1945, 1948; Stein and Hadziioannou, 1984; *web.mst.edu*; *www.ias.ac.in*; *www.wyatt.com*) such as that of Figure 1.2 can then be used to obtain the values for both M_w and R_g.

Performing a Zimm analysis on a single concentration is referred to as a partial Zimm analysis and is only valid

Figure 1.2 **Theoretical Zimm plot showing the double extrapolation of M_w as the circled point**

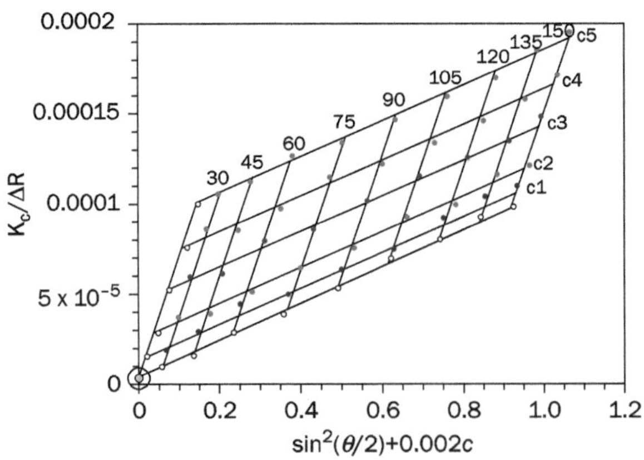

for dilute solutions of strong point scatterers. The partial Zimm does not, however, yield the second virial coefficient, due to the absence of the variable concentration of the sample.

In terms of an instrumentation system, the simplified configuration showing the key system elements is as shown in Figure 1.3. In operation, a metered sample would be injected into the light-scattering module using a micro-pump when the sampling process would be initiated. Once the measurements are completed, the sample is ejected from the chamber, which may then be flushed out with an appropriate fluid as required, before a new sample is injected. The quantities involved for the sample are of the order of a few micro-litres, requiring the use of precision systems to deliver metered quantities on demand (Chastek et al., 2007; Wyatt, 1993; Zhelev and Barudov, 2005).

Figure 1.3 Simplified system configuration for a light-scattering instrument

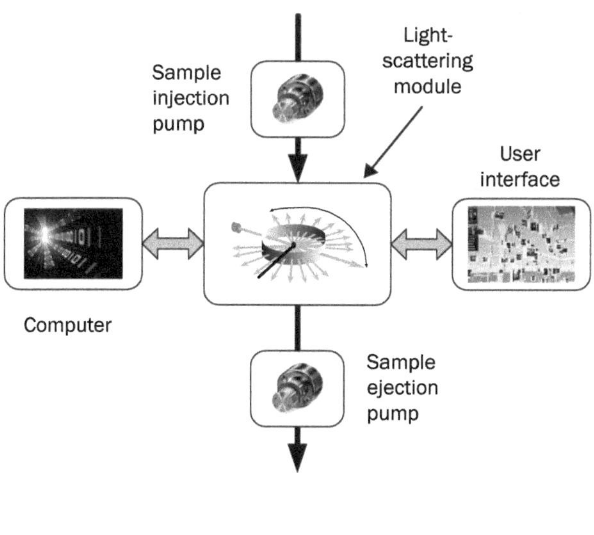

1.3 Innovation

However, in developing any complex product deploying a range of technologies, as is the case in the light-scattering system, solving the technological problems is often a relatively easy part of the overall development process, though the effort needed to provide effective and viable solutions must not of course be underestimated. In particular, once the need has been established, questions and issues such as the following need to be addressed:

- the robustness of the techniques to be deployed, expressed in terms such as accuracy, repeatability and means of validation;
- the influence of environmental effects on performance, as for instance the effect of variations in temperature, and the influence of factors such as humidity and pressure on performance;
- the control and operation strategies required and their links to the software implementation to be adopted;
- the nature of the user interface and the operator interaction with the system at all levels;
- the influence of external factors such as chemical resistance or use in the presence of ionising radiation;
- the need to 'future-proof' the system as far as possible, both in terms of the ability to introduce upgraded or enhanced components as well as software upgrades;
- the incorporation of embedded diagnostic procedures as an integral part of the system operation;
- the packaging requirements associated with the system such as size/mass limitations on use or target cost; also, will the system need to remain unaffected by vibration, shock and other issues during shipping or should it be

easily recalibrated at the user site without expert intervention following shipment?

Above all, however, the system must perform reliably according to specifications in whatever environment it is likely to be used in.

1.3.1 Innovation strategies

Innovation, and particularly the adoption of innovative approaches to the deployment and use of technology, is a key element of any product development strategy, with many businesses considering an ability to innovate the means of securing sustainable commercial advantage and leverage. However, wider societal interpretations may well be largely associated issues such as global stability and longevity.

In 1999, the then UK Department for Trade and Industry[1] defined innovation in what might be considered as relatively simplistic terms as:

> *. . . the successful exploitation of new ideas, products, materials, techniques and processes.*

This may be compared with the significantly more robust interpretation offered by the Council for Science and Technology as:

> *. . . the process by which ideas and knowledge are exploited for business purposes. It encompasses not only the creation of a new product, process or service but also the systems, processes, organisations, structures and all other aspects of a company's existing or future competitive edge such as distribution, marketing, branding and indeed the creation of a brand new*

> *market. The process draws on a range of intellectual and other inputs including knowledge of markets, customers, competitors, science, engineering and technology.* (CST, 2000)

Whatever the definition, it remains the case that different groups place different interpretations on the meaning and role of innovation. Thus it may be taken as representing:

- changes in organisational behaviour, activity or capacity;
- the enhancement of commercial acumen, markets or efficiencies;
- technological development, capture, absorption, verification, application, utilisation and other related factors.

These issues are reflected in the simplified model of the innovation process in Figure 1.4.

Additionally, issues of sustainability have led to the role of innovation being questioned. Consider therefore the view put forward by Charter and Clark that:

> *Sustainable innovation is a process where sustainability considerations (environmental, social, financial) are integrated into company systems from idea generation through to research and development (R&D) and commercialisation. This applies to products, services and technologies, as well as new business and organisational models.* (Charter and Clark, 2007)

In this approach to innovation, the emphasis is on the integration of organisational, architectural, socio-political and socio-technical issues within a sustainable context. These issues can in turn be reflected in the model of the innovation process proposed by Rothwell and shown in

Figure 1.4 The innovation process

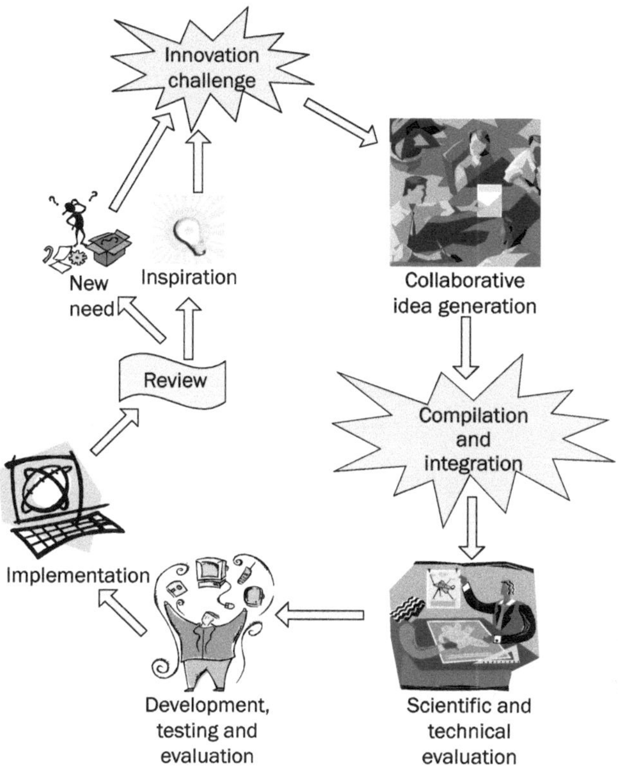

Figure 1.5, which tries to express the relationships between the various aspects of the innovation process and issues such as education and training, market factors, regulation and communications.

There is therefore a growing requirement both to drive the innovative process and to take account of societal change while doing so. However, it can also be argued that in order to develop and take forward the innovative process where profitability and growth are not the sole drivers, there is a need for a paradigm shift in approach to the innovation process itself.

Figure 1.5 Rothwell's innovation model

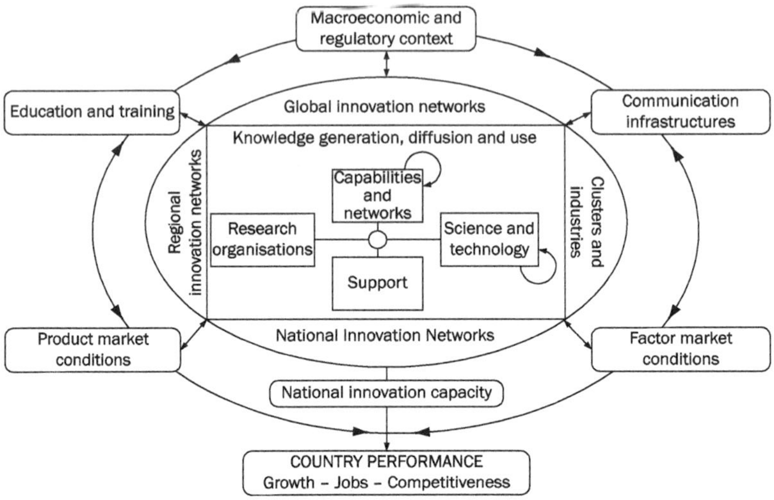

1.3.2 Open and closed innovation

Innovation, in all its forms, is key to the effective achievement of new generations of products and systems. However, in order to develop and take forward the innovative process to meet a new set of challenges, Christophers has suggested the need for a shift from the traditional approach, defined as *closed innovation*, with its orientation towards secrecy and the retention of ideas, to one of *open innovation* (Chesbrough, 2003; Chesbrough et al., 2006; *www.col-tech.org*) in which ideas and solutions are widely sought both from within and from outside the organisation.

The relationships between these two divergent approaches can be seen by reference to Table 1.1 and Figure 1.6. From these it can be seen that they each represent a significantly different focus on the innovation process, both in terms of the value of ideas and the ways in which such ideas are to be incorporated into that process. The revised methodology

Table 1.1 **Closed and open innovation characteristics and principles**

Characteristics of closed innovation	Characteristics of open innovation
▪ linear ▪ slow ▪ ideas represent a strategic advantage ▪ mentors ▪ learn by reverse engineering ▪ progress 'on the shoulders of giants' ▪ structured around the use of 'experts'	▪ networked ▪ fast ▪ ideas support development ▪ micromentors ▪ lessons learned benefit all ▪ progress through individual interaction with the many ▪ uses collective knowledge of groups, not individuals
Principles of closed innovation	**Principles of open innovation**
▪ The smart people in the field work for us.	▪ Not all the smart people in the field work for us. We need to work with smart people both inside and outside the company.
▪ To profit from R&D, we must discover it, develop it and ship it ourselves.	▪ External R&D can create significant value: internal R&D is needed to claim some portion of that value.
▪ If we discover it ourselves, we will get it to the market first.	▪ We don't have to originate the research to profit from it.
▪ If we create the most and the best ideas in the industry, we will win.	▪ If we make the best use of internal and external ideas, we will win.
▪ We should control our IP, so that our competitors don't profit from our ideas.	▪ We should profit from others' use of our IP, and we should buy others' IP whenever it advances our business model.

represented by open innovation has been adopted by organisations such as Proctor & Gamble (*secure3.verticali.net*) and the US Department of Education (*www.ed.gov*) to create platforms to develop and take forward new ideas, but perhaps more importantly to bring in new ways of thinking from outside the organisation.

Figure 1.6 **Closed and open innovation**

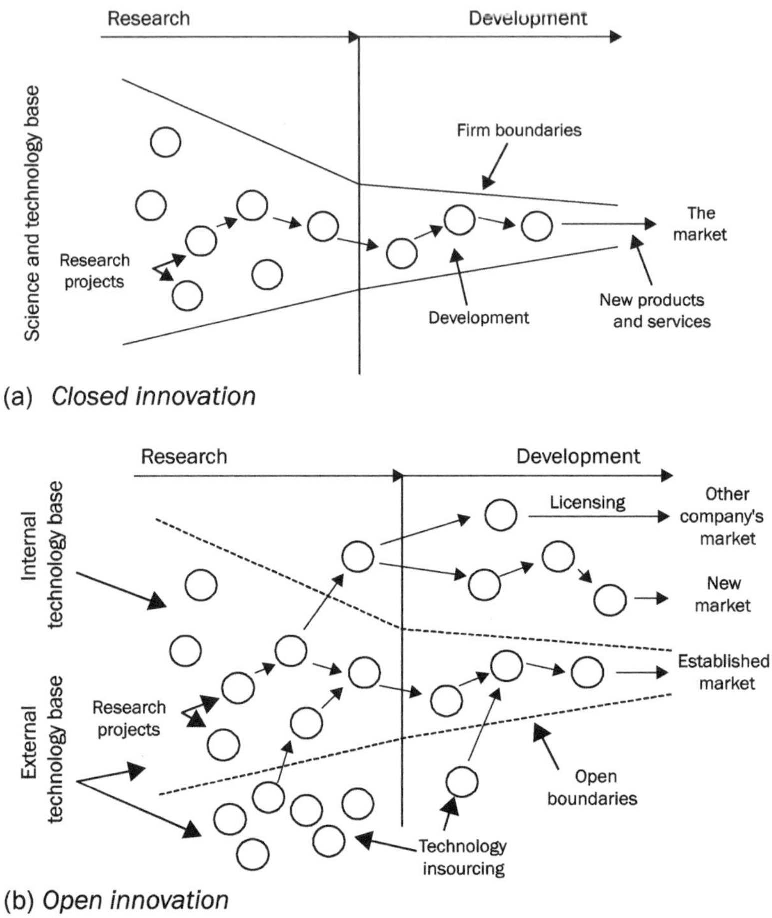

Though the underlying motivation, in one case profit and in the other an enhanced education system, may differ, both are exhibiting a degree of openness by inviting external bodies, groups and individuals to submit their ideas into a central 'pot'.

1.3.3 Managing complexity

In recent years, products and systems of all types from domestic appliances to vehicles have become increasingly more complex in respect of the achieved levels of integration of both mechanical and electrical systems with software to provide the increased, and increasing, levels of functionality. Thus a games console may use accelerometers or vision to record motion which is then translated into an on-screen response while a modern car will integrate multiple systems ranging from engine management to driver and passenger comfort controls, and potentially autonomous navigation (Thrun et al., 2006; Isermann, 2008).

There is, however, a danger that the developer fails to properly identify and recognise the links between the various system elements, and in particular the role of software in defining both system complexity and behaviour. Such a failure to properly recognise and understand the role of software led to the view in some quarters that '*it is all software now*', and that as changes to software were viewed as relatively easy, changes to the system were similarly easy. The result was that system software was often in a state of continuous change as '*it was simpler to make the change there*'! Both of these approaches have significant potential for delays and errors.

For instance, consider a flight of F22 *Raptor*s that were transferring from Hawaii to Okinawa in 1997. On crossing the International Date Line they lost all navigation aids and had to be escorted back to Hawaii. The issue was a coding error which resulted in an infinite loop as a consequence of the unexpected date change. In the words of Donald Shepperd, a former head of the US Air National Guard:

> *Reliance on electronics has changed the flight-test process. It used to be tails falling off, now it's typos that ground a fighter.* (*www.defenseindustrydaily.com*; *www.newscientist.com*)

Also, as systems complexity increases, it is difficult, and often indeed impossible, for any single individual to manage all the levels of detail required. This results in a partitioning of function to levels at which an individual can operate, demanding an understanding by the 'domain expert' of their specific task and its place in the wider overall system context. Thus the Mars Climate Orbiter (MCO) was lost in 1999 when a failure in communications between those responsible for navigation, who customarily used Newton-seconds to express thrust, and those responsible for the propulsion system, who in contrast used pound-seconds to express thrust, resulted in the MCO being 100km too close to Mars when it attempted to enter orbit (*www5.in.tum.de*).

1.4 Do the right thing

Despite an emphasis on the need to understand the customer when developing a product concept, there still too often remains an underlying assumption that the '*designer knows better*'. Further, many organisations still place the funding of the design process as a general business overhead, without any realisation of the fact that it is generally at the design stage that the most value is added and costs determined.

The understanding of the relationship between the establishment of need, effective and innovative design and product success has been established over many years (Andreasen and Hein, 1987), as has the relationship between the allocation of resources and spend, shown in simplified form in Figure 1.7, while the cost of error remediation at different stages of the development process is as suggested by Figure 1.8. Further, the nature of these relationships must be associated with a proper understanding, and management, of risk, as suggested by Figure 1.9.

Figure 1.7 Cost allocation and expenditure in product development

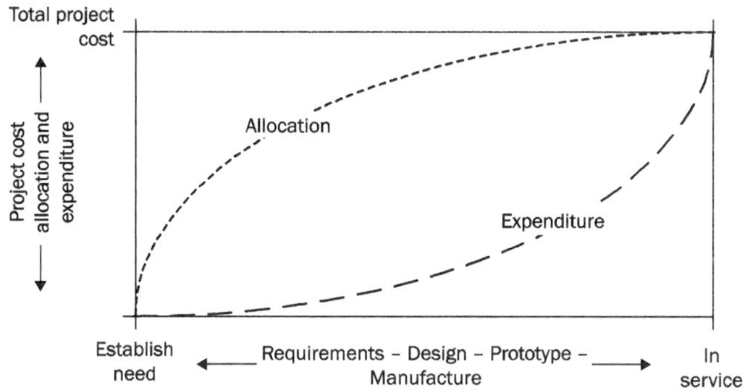

	% of total cost (cumulative)	
Product development phase	**Incurred**	**Committed**
Conceptual design	3–5	40–60
Design embodiment	5–8	60–80
Testing	8–10	80–90
Process planning	10–15	90–95
Production	15–100	95–100

Figure 1.8 Cost of error remediation in product development

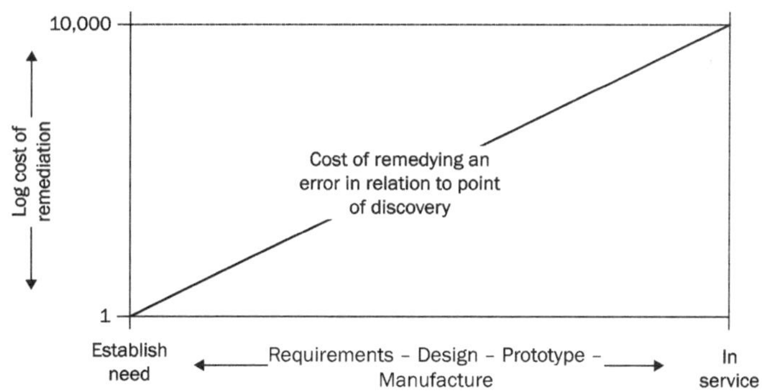

Error remediation	
Time change is made	**Relative cost**
During design	1
During testing	10
During process planning	100
During pilot production	1,000
During final production	10,000

Figure 1.9 The management of risk

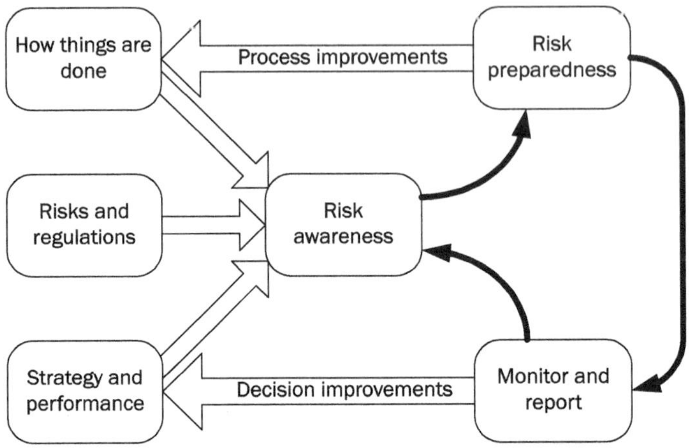

It is worthy of note that in many instances, engagement by senior management with the product development process tends to be high in the initial phases of the development process once the concept has been 'sold' and the project team set up. After these initial phases there is often a falling off of involvement until the point at which significant expenditure begins at the prototype stage, by which time it is possibly too late to influence matters!

Once the requirements are in place, the initial specification can be created and validated through further discussions with the target users in order to put the design specification against which the system is to be developed in place. There is also a requirement that the correct tools are used to support the various stages of the product development process. Figure 1.10 sets out and suggests the nature of these tools and their role in the product development process, including that of supporting effective communication between the various participants and stakeholders.

Figure 1.10 The product development process and tools to support that process (Wodehouse and Bradley, 2006)

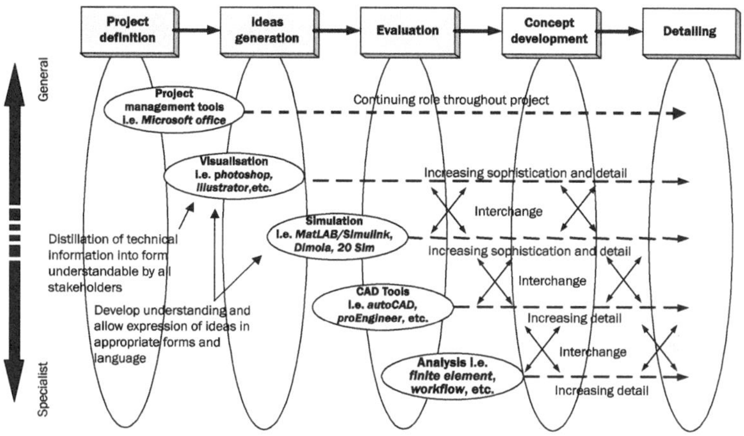

1.4.1 The design process

A proper understanding of the design process is therefore a crucial element in achieving innovation and ensuring that the resulting system meets all its functional requirements. This in turn implies that the design process itself takes account of and encompasses the following issues.

Understand the user

It can be argued that while satisfying the user is key to achieving a successful product or system, too often the views of the user are not properly sought out, identified and understood. This implies that the requirements capture process[2] needs to be both comprehensive and robust and supported by appropriate tools and methods and that approaches, as for instance those of user-centred design (Abras et al., 2004; Gulliksen et al., 2003; Vredenburg et al., 2002) as suggested by Figure 1.11, are deployed as required to ensure the engagement of the user.

Figure 1.11 User-centred design flow

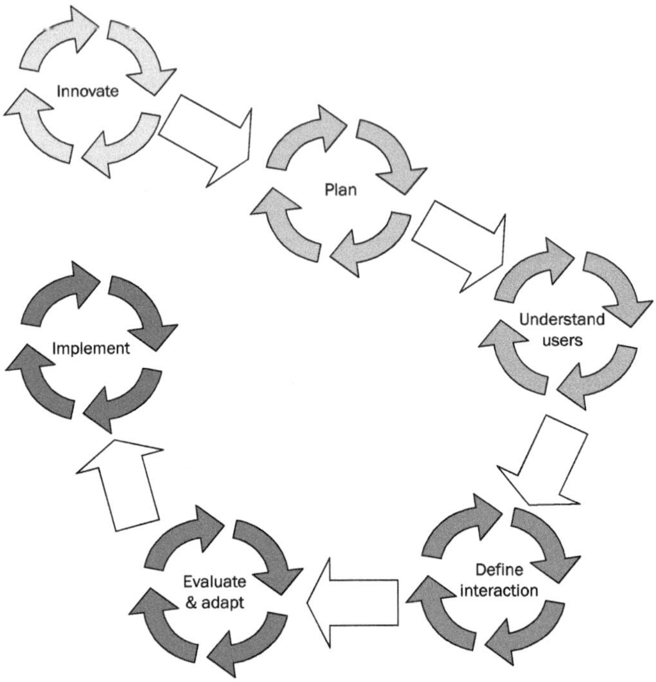

The approach adopted should, as far as possible, be generic

In a complex product the aim is to bring together and integrate a range of technologies to achieve the required outcomes. An over-concentration on a particular aspect through an inappropriate choice of method can at best be misleading, and at worst result in something which does not conform with the stated specifications.

Tools such as the V-model of Figure 1.12 can be used to support thinking, and in particular to identify and integrate the testing processes required to establish performance and functionality. It should, however, be recognised that any such tool provides only an approximation of the design process

Figure 1.12 **Simplified V-model for design management**

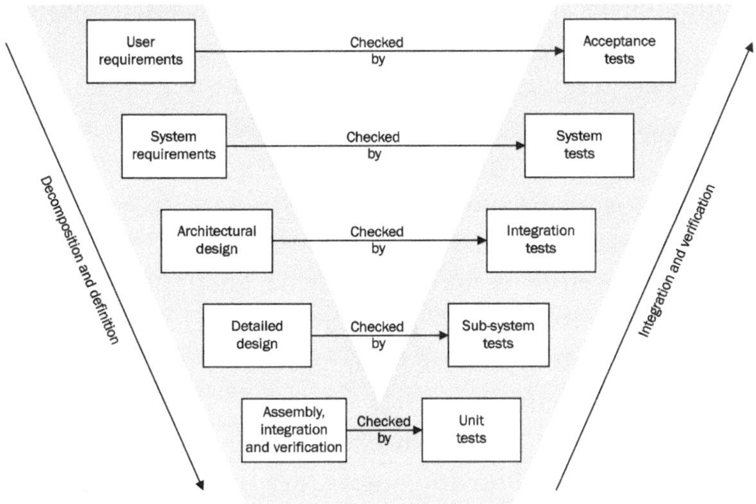

and must not be treated or considered as an absolute in determining either the progression of the design or the outcomes.

Search for and identify optimum solutions

This in turn requires that consideration is given to the outcomes and how these are to be met. It should also be recognised that there is not necessarily a single choice of optima that can be made, but that this may well be determined by context. Thus, it may be that the system is performance-driven, in which case this may set the lead criterion for any optimisation. Alternatively, it may be price that is the determining factor, in which case this would become the leading criterion. As with any optimisation process, there will be the need for trade-offs between different areas, as well as the need to avoid local minima on the way to achieving global minima.

Encourage creativity and invention

Continuing questions in product development and design are:

Where does new knowledge come from?

and

What is invention?

There is therefore a need to provide an opportunity for thinking unconstrained by previous examples along with an appropriate review process to support the evaluation of the proposed solutions.

Perhaps the best known illustration of this approach is the development of 'Post-It' notes (Fry and Silver, 2010; Petrovski, 1997) and masking tape (*web.mit.edu/invent/*) at 3M, which gives relevant employees some 15–20% of free time to allow them to work on their own projects. If the project is successful, it may then be spun off into a new business and the employee who originated the concept given an equity share. Indeed, many of its key products came from this free time. Other companies such as Google and IBM also offer their engineers and designers time to independently pursue ideas, and it is interesting to note that this encourages a more long-term view by removing the need to continually focus on short-term issues and problems (Bessant et al., 2009; London Business School, 2007; Grierson et al., 2004).

Ensure compatibility between system elements

In dealing with complexity there is, as has already been suggested, a tendency to have an undue focus on a particular aspect of the design to the exclusion of compatibility issues, as for instance the F22 *Raptor* and the transition across the International Date Line.

Minimise reliance on chance

Despite, or perhaps because of, the increasing ease of access to large volumes of information through the Internet, there will be increasingly large areas of knowledge which will remain inaccessible to a single individual by virtue of volume. It is therefore essential to establish means for sharing knowledge in relation to design tasks and problem-solving.

Facilitate the application of existing solutions

With increased mobility of staff, corporate memory is tending to shorten rapidly. Indeed, it has been said that with development of a new aircraft from first requirements specification to flight certification now taking on average 15 years, it is unlikely that anyone having a technical role at the commencement of the project will have any connection with that role at the end, even if still part of the organisation.

Tools to enable design recall and re-use are therefore essential if the benefits of prior work are not to be lost. This is of particular importance where a previous idea or concept may have been rejected or held in abeyance because the currently available technologies meant that at the time it was originally conceived it was either uneconomic or infeasible, or both, to implement.

Be compatible with information systems technologies

The design process is essentially structured around a search of the defined solution spaces to identify established practices and design embodiments and of finding ways of combining these to produce a new or improved product or outcome. Increasingly, methods based around the use of expert systems and structured knowledge bases are providing the means by which the initial searches can be made. It is however important to recognise that the output from these searches should be treated, in the first instance at least, more as suggestions or advice than absolutes (El-Nakla and Bradley, 2008).

Be easy to use

It is unrealistic to expect design engineers to operate in the world of the AI specialist. The inference and search techniques employed must therefore be transparent to the user with a user interface which allows inputs to be made and outputs presented in familiar language and formats.

Ensure objective evaluation

If the aim is to make 'right first time' a fully achievable goal, fully objective criteria for evaluating performance and outcomes must be set out in the statement of requirements. These then need to be tied in to appropriate testing procedures aimed at establishing functionality and conformance, as for instance is implied by the V-model format.

Reflect management thinking

Time constraints on the design process in highly competitive environments resulted in a shift from hierarchical to matrix management structures, but placed increased pressures on functional specialists to broaden their range of activity. One of the problems with the matrix organisation of Figure 1.13 is that it increases the level of competition for specialist design support and expertise (El-Nakla and Bradley, 2008), the resolution of which may occupy much time at director level. In contrast, the hierarchical organisation of Figure 1.14 is structured so that an individual project team contains more of the detailed technical resource required to complete its task.

Figure 1.13 Functional matrix project management structure

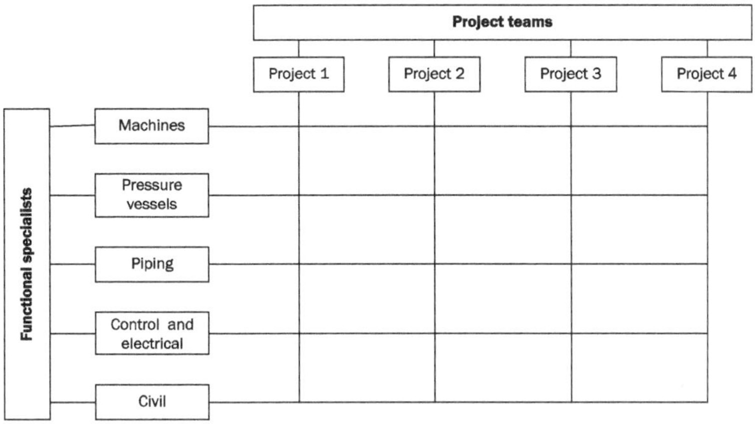

Figure 1.14 Hierarchical project management structure

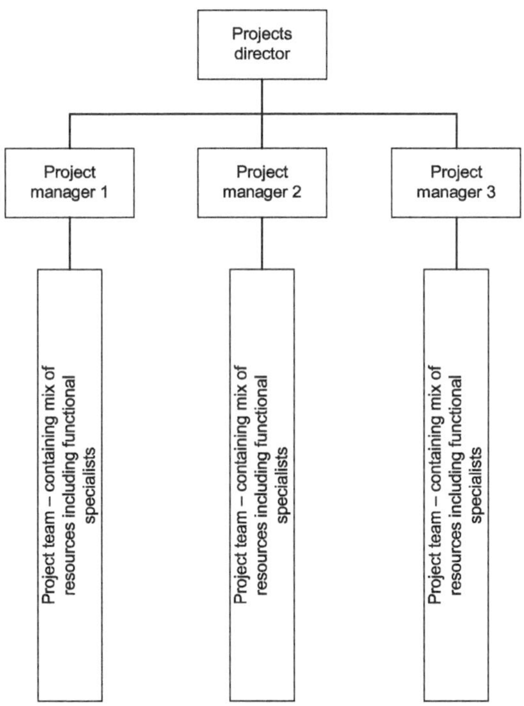

1.4.2 Requirements capture and analysis

An important, but often misunderstood, part of any design and innovation process is the need to properly understand and interpret customer needs and requirements. Any failure to do so will in general lead to an unsatisfactory product or system, irrespective of the excellence of its design or manufacture. Also in practice, errors associated with such a failure will typically not be identified until the product or system is with the customer, resulting in their further dissatisfaction.

This in turn leads to the question as to why errors are committed in requirements capture and analysis. In general it would be as a result of one or more of the following:

- hidden requirements or requirements that are perceived as 'obvious' by the customer and hence are not verbalised or otherwise expressed;
- 'requirement inflation' where the customer over-specifies the requirements for a given task to provide a perceived safety margin;
- the requirements as expressed by the customer are ambiguous;
- there is a lack of understanding of, and hence consideration of, whole life-cycle issues;
- the link between specification and appropriate system cost to the end customer has not been made.

In any design process, the best compromise must always be sought and it is the role of the development team to properly determine customer requirements. The development team therefore:

- establishes the need for a product;
- defines the market sector and demand;

- establishes customers' requirements, including the identification of any unstated expectations or bias;
- communicates customer requirements clearly and accurately in the form of the specification.

The specification should therefore set out and define:

- the function or purpose of the product or system;
- all requisite and required performance characteristics, including environmental and usage conditions, reliability, disposal, maintenance requirement and cost;
- aesthetic characteristics and human factors;
- all applicable standards and statutory regulations;
- packaging;
- quality requirements;
- the target market price for the finished system;
- if the design specification does not reflect the customers' requirements, the *wrong* system will be designed;
- if the design specification is ambiguous, the design *will not* meet the intended customer requirements.

1.5 Funding

For the majority of developers of a new technology or system, and in particular those outside the mainstream of large corporations, a crucial element of the development plan becomes that of '*who will pay for it?*'

With development capital in short supply and the venture capital (VC) market not necessarily willing to invest where there is a limited long-term potential for an initial placement offer (IPO) exit to the stock market and relatively low capital needs, the developer must be aware of the financial market that their innovation will be needing support from – even down to the best geographical locations for situating a company.

While the venture capital market is often perceived to be the best route forward, there are issues that should be taken into account when considering this option:

- In order to attract venture capital funding, it is often necessary to advance an overambitious development plan that requires far more money than is actually needed to launch the product. Aside from the issues of losing control of the business at too early a stage, such plans often lead to overruns in both costs and time to complete the market introduction of the product.

  Far worse, however, is the tendency to delay launch in order to complete '*just one more enhancement*', perhaps resulting in a product that is over-engineered, often for the wrong market, and with no funds left in reserve to make changes in response to market demands.
- Venture capital funding is normally based on excessive formal intellectual property (IP) portfolio development, which can increase costs, delay innovation and product launches in highly time-sensitive markets, as well as provide aggressive competitors with advance warning of new innovations coupled with the blueprints to copy them.
- When introducing a potentially disruptive technology that could result in a step-change with respect to existing technologies – the main driver for success of the product in the market – there is a strong push from the financiers to dilute this benefit by selling the advantages of the technology so that far larger sales margins can be achieved at lower volumes. This makes it difficult to sell, and easy for competitors to counter, and is a major cause of innovations failing to achieve their potential.

It may well therefore be better to develop a plan that leads to the design of products that, in the first instance, could be sold as original equipment manufacturer (OEM) components. Manufacturers are increasingly aware that innovation comes more rapidly, and at much lower cost, from small, entrepreneur-led organisations and it is becoming quite common for larger corporations to fund the creation of new products and companies to feed their own needs as part of a clustering strategy. With this in mind, R&D location as well as the likelihood of a new innovation cannibalising profitable parts of a target partner's income stream must be taken into account during the planning of a commercialisation strategy.

An alternative approach in the case of analytical research systems would possibly be to work with a large end-user laboratory to develop a specific solution to a generic analytical problem. If the solution is able to demonstrate substantial savings in the overall costs of drug discovery – such costs can reach sums of $1m/day or more– innovations can be funded without the innovator needing to become locked in to that investor. This is often a fast-track method for developing partnerships with major instrument manufacturers who, if all goes well, will be looking to purchase the product lines and/or company in a relatively short period of time providing a simple exit route for investors and stakeholders.

1.6 The prototype

Once the original concept is in place and validated against customer need and the requisite funding put in place, this needs to be supported by a working prototype that meets the established criteria in the development laboratory. When the prototype is available, it is essential to move swiftly to test the design in the market. This is in part to avoid the danger of over-engineering the design at an early stage. Such over-engineering is a characteristic of designers/inventors who can always see means of improving on an idea. However, this may well lead to delays in commercialisation and/or missing the target market and preventing use in fresh applications that could have been viable before such 'optimisation'. Indeed, referring to Figure 1.15, it can be seen that delays in development can easily lead to a loss of sales as competitors enter a market which has been prepared for them by the market initiators (Moore, 1991).

Figure 1.15 **Profiles of technology adoption and market penetration**

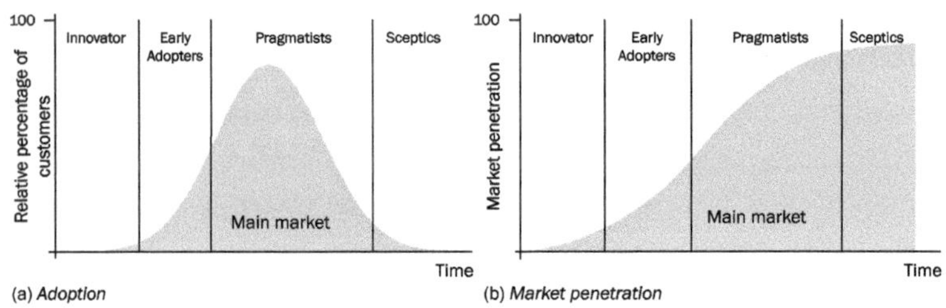

1.6.1 Beta test

Once the prototype concept has been validated, a beta test version can then be produced and made available to selected users. This stage has the primary aim of ironing out technical bugs and correcting issues relating to features such as the user interface and documentation based on questions raised when users attempt to put the system to use.

The beta test process should not be used to make major enhancements based on user feedback before launching V1.0, but to remove obvious problems which may detract from user acceptance. The aim should be to collect market and user feedback from at least six months of field use with paying customers and then to merge all the information gathered into the designs for V2.0. A good example of this approach is the launch of the Apple iPad, where V1.0 was used to create a new market despite obvious limitations such as the lack of an integral camera, then just as competitors appeared, V2.0 was announced to fix key issues identified by users that would have restricted future sales growth (Hernandez and GuoHua, 2011; Chou, 2010).

In selecting the beta test sites, a range of sites from different target application areas and geographical regions is required to ensure that variations in modes of use across industries and regions can either be accommodated into the revised design, or to restrict sales to those regions and applications that will work well with the V1.0 unit.

1.7 How to keep ahead of the competition – design in saleability!

In terms of product development, it is necessary to avoid the temptation to enhance the declared specification prior to release – thus giving an instrument a higher performance than that required to develop or expand the market. This simply reduces the long-term market impact of the new technology while giving competitors a technological benchmark to aim for, or indeed to copy, reducing opportunities to keep ahead of such competitors. Even if such enhancements would demonstrably open new markets, the temptation to divert to these new opportunities should be resisted as the result would be delay, increased cost, a loss of competitive advantage and a dislocation in the relationship with key specifying users. It can also act to put off those potential users, the pragmatists of Figure 1.15, who tend to be sceptical of performance claims, particularly when these are significantly ahead of what is currently seen as desirable or achievable without a shift to new, disruptive technologies.

The implication is therefore that it is necessary to keep some performance and capability in reserve, but not least to reduce the support issues following launch that are inevitable when operating new technological solutions at an innovation edge. This may mean limiting the stated performance of the V1.0 unit to a level that gives a user a significant benefit over what is currently being used such that when the competition catches up and attempts to re-establish market share, the specification can be opened up, typically via software enhancements, and sold as a revised version (V1.1, etc.) without the need for significant additional research and development. It also means that as users become aware of the improved capability to carry out tasks provided by the V1.0 unit, requests for enhanced performance that may result can be met while feeding the information received into a '*next generation*' (V2.0) system (Reppel et al., 2006). This approach also has the added advantage that the system is never operating at an edge of efficiency early in the product's life-cycle, allowing long-term issues to be dealt with as applications evolve.

In this context, it is also essential to build in regulatory compliance wherever possible. This is particularly important in markets such as the bio-pharmaceutical industry, where the auditability requirements of regulatory agencies such as the Food and Drug Administration (FDA) in the US or the European Medicines Authority (EMA) will need to be taken into account for many applications. Forward planning of this form can result in a market edge over more technology-focused competitors.

Other factors to be considered in relation to design for saleability are:

- **End of life** – recent requirements in many countries, as for instance the WEEE directive in Europe (*ec.europa.eu*), for the recycling of systems at the end of their useable life, require that appropriate and low-cost means of achieving this are built in from the very beginning of the design process.
- **Upgrade** – it is also a good idea to build in to early-stage products the ability to easily upgrade as enhancements are developed to extend product lifetime and again reduce recycling costs.
- **Size and price ratio** – where miniaturisation is an issue, it is often better to make size reduction a gradual process to sustain a higher market price (the '*empty box*' approach used by desktop PC manufacturers), or look to significantly expand the market through a quantum shift in both size and price. Price is always assumed to be lower the smaller a product becomes, while performance gains are not valued in the same, somewhat illogical, way.
- **Manufacturability** – many high-cost analytical instruments that have been built around technical innovations tend to neglect the importance of designing in the ability to mass-produce sub-systems at low cost, with many key components handmade and aligned for optimum performance. Historically, it was more important that an instrument was capable of measuring parameters crucial to the pharmaceutical industry than it was of low cost to build and easy to use. It was accepted that PhD-level scientists would be employed to operate such equipment, and the total cost of ownership was linked to labour rather than capital expenditure costs.
- **Study current systems** – in an area such as the bio-pharmaceutical industry, regulation tends to make it expensive to shift to different testing formats because of

the additional costs associated with validation and related issues. Also, reporting commitments in regulatory environments such as those run by the Food and Drug Administration (FDA) in the US can lead to enhanced sensitivity being a negative issue for users as previously undetected trace impurities that are totally benign have to be investigated once seen. Often therefore the need is to design solutions that match current methodologies despite the enhanced efficiency that may be obtainable when starting from scratch, as for instance problems of acceptance of 'circular' laboratories based around the use of robots.

- **Psychology** – laboratory analysts have often built career security on their ability to drive systems and to interpret data. Modern systems are often seen as a threat by such individuals, while management are always keen to keep labour costs low by employing less highly qualified analysts for routine work. It is necessary to either focus applications on areas where the system enhances the data capture process while allowing the analyst to be better positioned to interpret the information (as for instance in the Human Genome project – *www.ornl.gov*), or else to build into the system design features that allow sales strategies to overcome bench-level hostility, as for instance a reduction in time spent on routine versus investigative work, and to allow budgetary-level managers to buy in.
- **Market dynamics** – large corporate competition can always maintain dominance over better technology through market muscle, so any small incomer needs to have designs that can be rapidly evolved and enhanced to avoid this threat by remaining nimble enough to dodge around the lumbering opposition. Original equipment manufacturer (OEM) potential should also be carefully explored as it can provide protection from a 'Big Brother' interested in

protecting the edge they get against their main competition through the adoption of their designs. Do not simply rely on patents since commercial drivers could make it economically viable for competitors to ignore them and to copy published designs to retain their edge, safe in the knowledge that a small to medium enterprise (SME) has limited legal resources to counter them.

1.8 The evolutionary history of the NS4910 protein aggregation monitor (PAM)

Unlike many new products developed by a new company, often formed as a spin-out from an academic institution that had devised a new technological approach to solving an analytical problem from its own research, the first designs for what has recently been commercially launched as the PAM came from a marketing company listening to customers and translating the feedback into a new concept for use in early-stage drug discovery research. It was quickly apparent that there were far more potential customers interested in using light-scattering for size-related measurements in their research than were actually intending to purchase. Anecdotal evidence suggested there could be as many as nine potential users for every one actual.

A number of critical issues came out of this extended listening exercise that needed to be addressed in designing next-generation tools if the latent demand identified was to be tapped:

- **Minimal sample use** – vast numbers of difficult-to-purify biological samples need screening during early drug discovery, which severely restricts the use of traditional light-scattering systems where sample volumes in excess of $10\mu l$ are required.

- **Simplicity** – traditional detectors have data-intensive outputs that require high-level (PhD) analysts to interpret the results, adding to analysis costs and workflow bottlenecks as expertise has to be concentrated into specialist facilities.
- **Cost** – light-scattering detectors have been developed over the years as a specialist product, where hand-built, precision optical systems could be sold at a premium price to specialist analytical groups.
- **Size** – many bio-pharmaceutical research labs have severe constraints on the amount of bench space that can be made available for traditional analysis systems that typically required a bench-top PC to run them in addition to the similarly bulky detector itself, coupled in many cases to an even larger sample purification/separation system. In many cases, instrument size was also dictated by a desire on the part of manufacturers to add a premium market price.

It is interesting to note that, with the possible exception of the first, none of these criteria related back to technological innovation but rather required an understanding of bio-pharmaceutical research lab workflows and the economics of drug development.

This analysis led to a recognition that there was a market opportunity for simple light-scattering tools that were designed to do specific applications rather than be used as a data-intensive, flexible analysis tool capable of being switched from one complex analysis problem to another. Indeed, by focusing on the development of compact, low-cost, targeted screening tools for use at the lab bench by research biochemists as an integral part of their experiments, the bottlenecks and quality issues that existed within analytical departments could also be alleviated by reducing the numbers of routine samples needing to be subjected to a full-blown analysis.

This realisation led to significant design changes becoming apparent for the detector that was under development at the time. Firstly, the actual theoretical approach used in the determination of size was changed to enable the overall cost and size (including sample volume) to be reduced. Secondly, it was decided to design the first detector around a specific application: screening protein solutions for the presence of aggregates prior to attempting to grow single crystals for x-ray crystallography. This application, while extremely specialised, was chosen in part because of the large numbers of protein structure labs around the world (more than 3,000) that would benefit from such screening once the detector (called *AggreKem*) was launched, while at the same time being too small a niche for larger competitors to bother about. It was also a perfect testing ground for the design model used and, if successful, would enable other niche applications centred around protein–protein interactions to be exploited with very little additional development expenditure.

The *AggreKem* as designed had a minimum sample volume of 5μl and used fibre-optic technology to keep the overall footprint small. Lastly, to avoid the need for a desktop PC, it was designed with a built-in Windows PC with custom software that enabled the aggregate content to be measured in 1 second using just three keystrokes from a custom interface that did away with the need for a computer keyboard (Figure 1.16). The overall package when launched sold for $16,000 against $30,000+ for other LS detectors.

Figure 1.16 Light-scattering evolution showing the traditional market leader (bottom), original *AggreKem* with built-in PC (top), and NS4910 PAM (front)

The *AggreKem* was very well received by the customer base when launched, but did have some key limitations that restricted sales:

1. Many customers wanted to use the detector to monitor aggregation kinetics as well as merely screen for their presence which, while possible, was extremely tedious using the single measurement *AggreKem* as results had to be noted down at timed intervals then entered into a spreadsheet.
2. The unit price was above the threshold used by most purchasing departments to denote whether a purchase was made from an operating budget under the direct control of a laboratory manager or had to be included in

the annual capital expenditure (cap-ex) budget and put out to competitive tender. In the latter case, it could take 18 months to go from interest to placing an order and ran the risk of competitors trying to sell up to one of their 'more sophisticated' systems to non-technical accounts personnel, whereas when bought by the lab manager the decision can be made on the spot by someone directly involved in the research who understands the relative merits of a screening tool and an analysis system.

3. The low-volume flow-path generated significant back pressure when injecting a sample because of the low ID tubing used, making it difficult to achieve the minimum sample volume reproducibly and necessitating increased sample volumes to be used – though these were still below other instruments available.

The *AggreKem* was also launched around the same time as the Netbook became available, reducing both the size and cost of laboratory computing to a level below that of the internal PC/LCD screen combo used in the original design.

Reacting to this experience, the detector was redesigned from the ground up to produce the *Norton PAM* (protein aggregation monitor). As can be seen in Figure 1.16, the PAM has done away with the internal PC and has been designed to fit inside a commercially available external hard drive casing to reduce size and cost. The earlier fibre-optic technology has been retained, but the flow-path/sample injection sub-system has been replaced with a simple port that enables the tip of a laboratory pipette to be inserted and the aggregates measured within the sampling tip – effectively a zero-volume measurement since the 2μl sample aliquot taken up in the pipette is never dispensed into the PAM and can be returned to a sample tube after analysis.

The software was rewritten in Java to make it possible to be used with the multiplicity of operating systems now available in the lab, and data is collected using a simple USB connection. Furthermore, the ability to automate kinetic measurements has also been added.

The net result is a detector that sells at a higher margin for half the price of an *AggreKem* despite the enhanced capabilities. This drop in overall price has placed it firmly into the '*operating*' rather than '*cap-ex*' budget class, accelerating sales growth.

1.9 Conclusions

The development of a complex product has been considered not only from the technical perspective but from consideration of user needs, financing, development strategy, market understanding and associated short-, medium- and long-term goals. The need for an innovative approach which brings together ideas from as wide a range and variety of sources as possible has been highlighted, along with the need to put in place strategies which encourage innovative thinking at all levels in an organisation. This open approach has, however, been contrasted with the need to impose constraints to avoid the 'overdevelopment' of the system in the early stages, both to support its effective introduction to the market and to keep something 'in reserve' to respond to market reaction, and competitors.

This approach has been illustrated by consideration of the evolution and development of first the *AggreKem* and then the *Norton NS4910 PAM* aggregation monitors for the pharmaceutical industry. Together, these illustrate the progression from an instrument intended for use by specialists to much smaller units which provide their functionality through the use of advanced computing strategies.

Throughout, it should be noted that product development is not a straightforward process focused on the resolution of technical issues, but is often a more personal process that requires both persistence and perseverance.

1.10 Notes

1. Remodelled first as the Department for Business, Enterprise and Regulatory Reform and then as the Department for Business, Innovation and Skills.
2. See 'Requirements capture and analysis', this chapter p. 26.

1.11 Acknowledgements

The authors would like to acknowledge the contribution to the content of the chapter through research, discussion, argument and debate over the years of numerous colleagues, research collaborators and students. In particular; David Dawson, Stuart Burge, John Millbank, Andrew Wodehouse and Samir El-Nakla provided effective sounding boards.

Also, Dr Ian Brindle, VP Research at Brock University who supported the development of the PAM prototype during the creation of Norton Scientific Inc. and Adrian Snell who facilitated a lexible approach to software development.

1.12 References

Abras, C., Maloney-Krichmar, D. and Preece, J. (2004) 'User-Centered Design', in *Berkshire Encyclopedia of Human–Computer Interaction*, edited by W.S. Bainbridge. Great Barrington, MA: Berkshire Publishing.

Ahmad, H., Imad, A. and Striegel, A.M. (2009) 'A Coupled MALS-DRI Method for Simultaneous Zimm and ∂n/∂c Plot Construction', *Instrumentation Science & Technology*, 37(5): 574–83.

Andreasen, M.M. and Hein, L. (1987) *Integrated Product Development*. Cambridge: IFS Publishers.

Bessant, J., Möslein, K. and von Stamm, B. (2009) 'In Search of Innovation – When companies try to come up with new ideas, they too often look only where they always look. That won't get them anywhere', *Wall Street Journal*, 22 June. Available at: *http://www.wiworkforce.com/articles/innovation-search.pdf* [accessed 8 February 2011].

Charter, M. and Clark, T. (2007) *Sustainable Innovation*. Guildford: University of Surrey, Centre for Sustainable Design.

Chastek, T.Q., Beers, K.L. and Amis, E.J. (2007) 'Miniaturized dynamic light scattering instrumentation for use in microfluidic applications', *Review of Scientific Instruments*, 78(7).

Chesbrough, H.W. (2003) *Open Innovation: the new imperative for creating and profiting from technology*. Boston: Harvard Business School Press.

Chesbrough, H.W., Vanhaverbeke, W. and West, J. (eds) (2006) *Open Innovation – Researching a New Paradigm*. Oxford: Oxford University Press.

Chou, W. (2010) 'Marketing Smart', *IT Professional*, 12(3): 8–10.

Council for Science and Technology. (2000) *Technology Matters: Report on the exploitation of science and technology by UK business*. London: Department for Trade and Industry.

ec.europa.eu/environment/waste/weee/index_en.htm [accessed 16 April 2010].

Einstein, A. (1910) 'Theory of Opalescence of Homogeneous Liquids and Mixtures of Liquids in the Vicinity of the Critical State', *Annals of Physics (Leipzig)*, 33:1275 (in *Colloid Chemistry: Vol. I*, edited by J. Alexander. New York: Rheinhold, 1913, p. 323).

El-Nakla, S. and Bradley, D.A. (2008) 'Case-Based Reasoning in the Design of Mechatronic Systems', Mechatronics 2008, Limerick.

Fry, A. and Silver, S. (2010) 'First Person: "We invented the Post-it Note" ', *Financial Times*. Available at: *www.ft.com/cms/s/2/f08e8a9a-fcd7-11df-ae2d-00144feab49a.html#axzz18hyDnyKX* [accessed 13 April 2011].

goldbook.iupac.org/Z06748.html [accessed 5 October 2010].

Grierson, H., Nicol, D., Littlejohn, A. and Wodehouse, A. (2004) 'Structuring and Sharing Information Resources to Support Concept Development and Design Learning', Networked Learning Conference 2004. Available at: *http://www.networkedlearningconference.org.uk/past/nlc2004/proceedings/individual_papers/grierson_et_al.htm* [accessed 8 February 2011].

Gulliksen, J., Göransson, B., Boivie, I., Blomkvist, S., Persson, J. and Cajander, Å. (2003) 'Key principles for user-centred systems design', *Behaviour & Information Technology*, 22(6): 397–409.

Hernandez, J.A. and GuoHua, Y. (2011) 'Is it just technology or other else behind Apple? Study of the marketing strategy for iPad launching as its new initiative product', Proceedings of the 2nd International Conference on Business and Economic Research (ICBER 2011): 2653.

Hoffman, M. and Beaumont, T. (1997) *Application Development: Managing the Project Life Cycle*. Ketchum, ID: MC Press.

Isermann, R. (2008) 'Mechatronic systems – Innovative products with embedded control', *Control Engineering Practice*, 16(1): 14–29.

Moore, G. (1991) *Crossing the chasm: marketing and selling high-tech products to mainstream customers*. New York: Harper Collins.

Petrovski, H. (1997) *The Evolution of Useful Things: How Everyday Artefacts – from Forks and Pins to Paperclips and Zippers – Came to be as They are*. New York: Vintage Books.

Reppel, A.E., Szmigin, I. and Gruber, T. (2006) 'The iPod phenomenon: identifying a market leader's secrets through qualitative marketing research', *Journal of Product & Brand Management*, 15(4): 239–49.

secure3.verticali.net/pg-connection-portal/ctx/noauth/PortalHome.do.

Stein, R.S. and Hadziioannou, G. (1984) 'Generalization of the Zimm equation for scattering from concentrated solutions', *Macromolecules*, 17(5): 1059–62.

Thrun, S. et al. (2006) 'Stanley: The Robot that won the DARPA Grand Challenge', in *The 2005 DARPA Grand Challenge: The Great Robot Race*, edited by M. Buehler, K. Iagnemma and S. Singh. Berlin: Springer.

Vredenburg, K., Ji-Ye, M., Smith, P.W. and Carey, T. (2002) 'A Survey of User-Centered Design Practice', Proceedings of SIGCHI Conference on Human Factors in Computing Systems: Changing our world, changing ourselves (CHI'02), CHI Letters, 4(1): 471–8.

web.mit.edu/invent/iow/drew.html [accessed 13 April 2011].

web.mst.edu/~wlf/mw/Zimm.html [accessed 5 October 2010].

Wodehouse, A. and Bradley, D.A. (2006) 'Gaming techniques and the product development process: commonalities & cross applications', *Journal of Design Research*, 5(2): 155–71.

www.col-tech.org/coltech/members-only/innovacion/THE%20ERA%20OF%20OPEN%20INNOVATION%20-%20SLOAN%20MANAGEMENT.pdf [accessed 5 October 2010].

www.defenseindustrydaily.com/f22-squadron-shot-down-by-the-international-date-line-03087/ [accessed 2 February 2010].

www.ed.gov/open/plan/flagship-initiative-collaboration.

www.ias.ac.in/initiat/sci_ed/resources/chemistry/LightScat.pdf [accessed 5 October 2010].

www.newscientist.com/article/mg19626359.900-the-doh-of-technology.html?page=2 [accessed 2 February 2010].

www.ornl.gov/sci/techresources/Human_Genome/home.shtml [accessed 16 April 2010].

www.wyatt.com/solutions/software/calypso-software.html [accessed 16 April 2010].

www5.in.tum.de/~huckle/bugse.html [accessed 20 January 2010].

Wyatt, P.J. (1993) 'Light scattering and the absolute characterization of macromolecules', *Analytica Chimica Acta*, 272(1): 1–40.

Zhelev, N. and Barudov, S. (2005) 'Laser light scattering applications in biotechnology', *Biotechnology & Biotechnology Equipment*, 19(3): 3–8.

Zimm, B.H. (1945) 'Molecular Theory of the Scattering of Light in Fluids', *Journal of Chemical Physics*, 13: 141–5.

Zimm, B.H. (1948) 'The Scattering of Light and the Radial Distribution Function of High Polymer Solutions', *Journal of Chemical Physics*, 16: 1063–9.

2

Planar micromanipulation on microconveyor platforms: recent developments

Panos Lazarou and Nikos A. Aspragathos

Abstract: Nowadays microcomponents are fabricated in very large quantities, whereas their automatic sorting, transportation and alignment in order to assemble the final MEMS products are still lacking in efficiency, thereby bottlenecking production. This chapter highlights recent developments in planar manipulation of microparts with the usage of microconveyor platforms instead of robotic grippers or automated manipulation systems. A classification of the various actuator types is made based on their input energy and a variety of existing platforms or promising implementations are presented. With certain design and performance characteristics as targets, the feasibility of these implementations for planar microconveyance is then examined and the most suitable ones are determined. Additionally, an integrated approach for manipulation on microconveyor platforms based on programmable force fields is presented, providing sensorless, automatic alignment and orientation of polygonal parts, verified by simulation results.

Key words: planar micromanipulation; microconveyor platforms; programmable force fields; MEMS.

2.1 Introduction

All MEMS systems, regardless of the kind of components they consist of, must be assembled. Current industrial microfabrication methods produce microcomponents in very large quantities; these components need to be properly sorted, transported and aligned for the assembly and packaging of the final MEMS product. Established solutions such as part-feeders and serial/automated-serial assembly with microgrippers are not efficient enough for the constantly increasing desire for mass production of smaller MEMS devices, because the sequential operations are not flexible enough, do not last long and present high cost.

Additionally, the integration of microscale modules, such as sensors, actuators and transceivers, as well as microfluidic, electromechanical and optoelectronic devices into heterogeneous, multifunctional systems in very large numbers is one of the big challenges in the field of MEMS production. The need for fast parallel procedures of production, transportation, sorting, positioning and orienting of microparts, without the usage of robotic microgrippers, micromanipulators or part-feeders, is becoming more and more evident. Recently, a variety of MEMS systems for the realisation of micro-locomotion in the form of microconveyor platforms have been presented and can potentially be part of the solution to the aforementioned handling and assembling challenges. Microconveyor platforms can be generally distinguished into two basic categories: contact and contact-free, depending on whether the platform's actuators come in contact with the moving object or not (Stemme and Ebefors, 2006; Ebefors et al., 2000).

In most realisations of platforms where the actuators are in contact with the moving object, arrays of moveable legs erected from the substrate surface are inducing motion. The legs actuation is based on various principles such as thermal, electrostatic and magnetic actuation. The activation of these actuators can be simultaneous (synchronous driving) or selective/controlled (asynchronous driving), which can usually provide higher speed and smoother motion.

In contact-free platforms, various force fields, such as electrostatic, magnetic or pneumatic, are used to create a cushion on which the part levitates. The forces of these fields need to be directionally dependent in order to drive the part's manipulation on the workplane (e.g. the directed air streams for a pneumatic system). The main advantage of the contact-free microconveyor platforms is low friction, while the drawback is their high sensitivity to the cushion thickness (load-dependent), which can be quite difficult to control.

A different classification of the various actuator types usable in microconveyance platforms can be made based on the type of their input energy according to Stemme and Ebefors (2006). In thermomechanical actuators, the input energy is thermal, while the output energy is mechanical and the efficiency of the energy conversion is determined according to the Carnot cycle. Most of the microconveyor platforms have been developed following this conversion scheme. There are two main types of actuators in this area: (a) thermal actuators, where different thermal expansion coefficients of two thin laminas cause bending of the composite structure upon heating and cooling; and (b) shape memory alloy actuators, where the input thermal energy triggers a phase transition in the alloy, resulting in the shape recovery of a previously deformed state.

In electrostatic actuators, the operating principle is based on electrostatic attractive and repulsive forces between electrical charges. Indicative actuator types are electrostatic actuators with gas or vacuum as dielectric isolation between electrodes. Similarly, in electromechanical actuators the input electrical energy is transformed to mechanical energy. Indicative types of actuators are: (a) electromagnetic actuators, where the Lorentz interaction between a moving electrical charge and a magnetic field is exploited to supply either translational or rotational mechanical energy to the conductor; and (b) piezoelectric actuators, where the piezoelectric effect resulting from the interaction of an imposed electric field and electrical dipoles in the material results in a deformation. In the case of fluidic-mechanical actuators (mainly pneumatic) the pressure of a fluid is converted into mechanical energy, either rotational or translational. Although sometimes considered as a 'traditional actuator' this transducer finds remarkable applications in MEMS and especially in microconveyor platforms.

In addition to the various implementations of microconveyor platforms, several manipulation strategies of parts on such platforms have been proposed. These strategies rely on a variety of programmable force fields, in other words mathematical expressions of force fields produced by functions of potential or by other methods. The characterisation 'programmable' comes from the fact that most of them can be diffused/applied on a microconveyor platform by proper programming of the actuator cells, so that they induce motion as directed by the diffused field. The main attractive feature is that the majority of these fields can offer sorting, two-dimensional translation or rotation on a plane with little or no need for sensors or imaging systems.

The rest of this chapter is divided into the following sections: in the first section, indicative implementations of microconveyor actuator platforms are presented, according to the classification. Subsequently, the design and performance characteristics of these implementations are compared and their suitability for planar manipulation is determined. The second section includes a brief introduction of indicative types of programmable force fields, as well as an integrated approach for field design and programming of microconveyor platforms for sensorless planar micromanipulation. The chapter concludes with future research directions and proposals.

2.2 Microconveyor platforms for micromanipulation

As previously mentioned, manipulation of microparts and microcomponents is a very important issue during the assembly and packaging of MEMS devices. Planar actuator microconveyor platforms can provide an alternative towards handling, sorting and alignment but they also need to have certain design and operational characteristics mainly at actuator level. First of all, it is crucial for such micromotion systems to realise enough load capacity in order to move and position the targeted parts, meaning high forces and torques. It is therefore essential to have actuators able to generate forces not only to lift a part out of the plane in order to avoid surface sticking but to also move it in the in-plane direction. Additionally, long strokes (for contact systems) or long motion steps (for non-contact systems) are desirable, as well as reasonable conveyance speed. Power consumption, dimensions and the complexity of fabrication and operation of actuators and platform also need to be taken into consideration. Usually, there is a trade-off between the above-mentioned factors but the main focus should be given to large strokes and high forces, according to Ebefors (2000) and Fearing (1998). The speed can be of less importance as long as the actuation frequency is reasonably high, i.e. in the range of a couple of Hz up to thousands of Hz.

This section presents indicative implementations of actuator microconveyor platforms or actuators that could potentially be used for planar micromanipulation, separated into categories according to the classification made in the introduction: thermomechanical, electrostatic, electromechanical and fluidic-mechanical. For each implementation, a brief description of the actuators, their operation principle as well as their reported performance characteristics is given; unfortunately not all the information relevant within the scope of this chapter has been reported by the respective authors. At the end of the section, a comparison between the most promising actuators or platforms for planar micromanipulation is made in a concentrative table and the most suitable of them are determined based on the criteria proposed in the previous paragraph.

2.2.1 Thermomechanical

Thermomechanical systems use the physical expansion or contraction that occurs in materials as they undergo temperature changes. Such thermal changes result from the conduction of heat energy into a material and may occur over a wide range of speeds. Thermomechanical actuators generally require the removal of heat energy to re-establish the previous condition (reverse transformation), which is probably their most important limitation. Overall performance is highly dependent on surrounding temperature and heat dissipation conditions, as well as strain level in the material and strain history. Some of these actuators rely on significant currents and/or power for their operation, which may require additional, forced cooling systems. It should also be mentioned that all actuators of this type require contact with the manipulated part.

Thermal bimorph (cilia)

Thermal bimorph actuators consist of deformable microstructures that curl into and out of the substrate plane. The curling of the cantilever actuators occurs due to the different coefficients of thermal expansion of two layers from materials such as polyimide, which make up the structures. When an electric current is passed through a heater resistor sandwiched between the two layers, the temperature of the actuator increases, and the structure (initially deflected out-of-plane) deflects downward, producing both horizontal and vertical displacements.

In the platform proposed by Suh et al. (1997), the actuators have triangle shapes and are grouped into 1x1 mm2 cells ('motion pixels'); each cell contains four orthogonally arranged cilia actuators (Figure 2.1). The platform was designed to perform simple array manipulations such as linear and diagonal translations as well as more complex vector-field manipulations such as rotation, alignment and centring by proper coordination of the deflections of the actuators. All of these operations were successfully performed. The vertical and the horizontal displacements of the cilia tips are 95μm and 17μm respectively, the lifting capacity is 23μN, the maximum achieved horizontal velocity is 1mm/s and the required power is 35mW per actuator.

In Ataka et al. (2010), a platform of stack-integrated arrays of microactuators and sensors that provide planar micropart motion has been proposed. In this system, there are four orthogonal actuators in each cell, as shown in

Figure 2.1 Group of actuators and one motion pixel cell of the cilia platform (Suh et al., 1997)

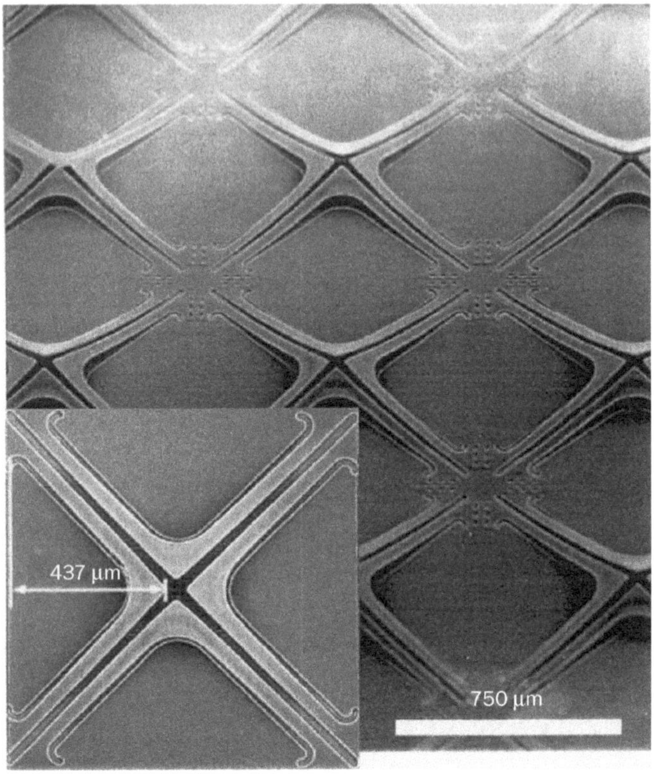

Figure 2.2. Part rotation and translation are performed by feedback control on the actuators, based on the information provided by the sensors. Experimental results show that a 180-degree rotation of an object can be performed in 33 seconds with feedback data, with a driving voltage of 19V and an operational frequency of 16Hz. Successful straight and orthogonal conveyance, automatic alignment as well as simultaneous conveyance of multiple objects have been demonstrated by the authors (Ataka et al., 2007a, 2007b).

Another implementation of a cilia platform is introduced in Wu and Hsu (2006). In this publication, an electrothermally

Figure 2.2 A work cell of the planar cilia platform (Ataka et al., 2010; © [2010] IEEE)

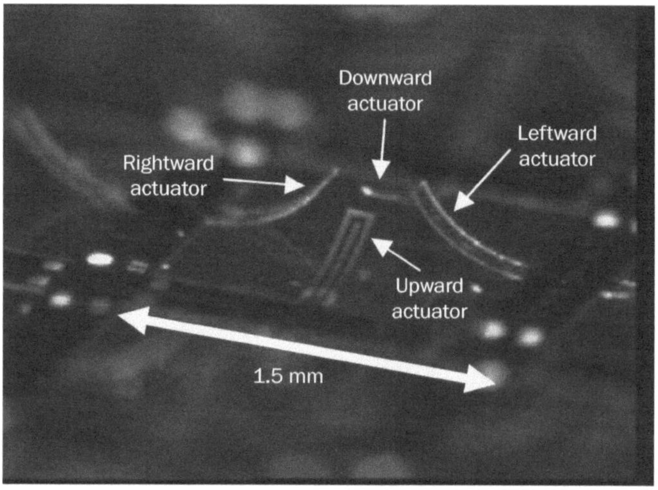

driven microactuator with adjustable height is proposed. The microactuator is based on the principle of thermal bimorph actuation with two long conveying fingers to exert out-of-plane bending motions in the transversal direction, which are connected and lifted by an initially curved height adjuster in the longitudinal direction. The platform can provide conveyance of micro-objects between two plane levels of different heights. The actuator of dimensions 400x50x4.5μm can provide 5μm vertical displacements by the height adjuster at 1V and 18μm lateral displacements by the conveying finger at 2V. The approximate maximum load per actuator is about 0.196mg and the maximum velocity for conveyance is about 1.0–1.5μm/s at an operating frequency of 5Hz.

Polyimide V-groove joint

The polyimide V-groove joint actuator has been developed for walking microrobot applications but also for use in microconveyor platforms (Ebefors et al., 2000; Ebefors, 2000). The principle of a 500x600x30μm polyimide joint actuator is shown in Figure 2.3. The out-of-plane element is rotated using thermal shrinking of the polyimide in the V-grooves, where polysilicon or metal heaters are integrated. By driving a current through these heaters, local heating is achieved in the joint, resulting in thermal expansion of the polyimide. The V shape of the joint results in larger absolute expansion at the top of the V-groove than at the bottom and thus the displacement Δx is achieved. The maximum measured conveyance speed of this actuator is 12mm/s with an applied voltage of 23V and 250Hz frequency, exerting 100mN. The payload per actuator is 350mg and the observed displacement is 170μm with an actuation frequency of 3Hz and a power consumption of 175mW.

Figure 2.3 **Schematic of actuation principle based on a four V-groove joint. By heating the joint a horizontal displacement Δx is obtained due to greater thermal expansion of the polyimide at the top of the V-groove than at the bottom (adapted from Ebefors, 2000)**

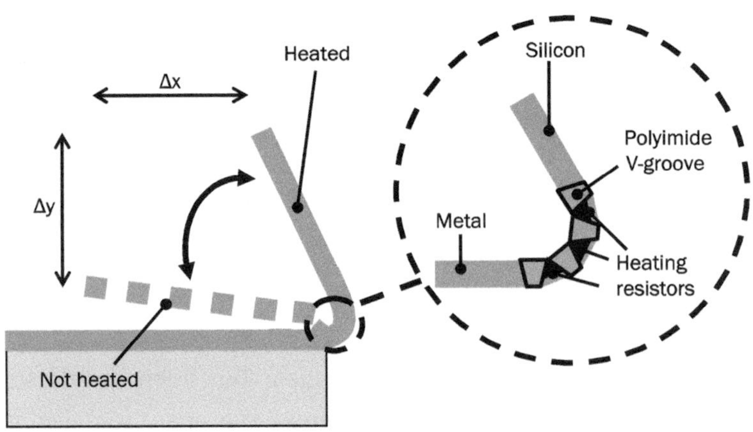

Thermally compliant

Another category of thermal actuators comprises single material structures instead of double layers as in the bimaterial thermal actuators these are the thermally compliant ones. The common name of this class of actuators is 'electro-thermally actuated compliant' (ETC). The application of a voltage difference between two points of electrically conducting and elastic continuum produces Joule heating. The topology and shape of this continuum gives rise to non-uniform Joule heating and hence non-uniform thermal expansion.

The schematic of one of the first ETC microactuators introduced in Guckel et al. (1992) is shown in Figure 2.4. The thin arm (hot-beam) has a higher electrical resistance than the wide arm (cold-beam), gets heated more and consequently elongates more. Similar design models have

Figure 2.4 **Schematic of an ECT actuator (adapted from Guckel et al., 1992)**

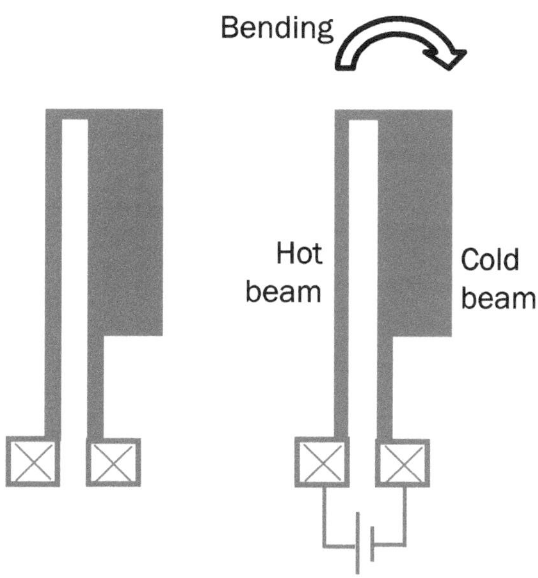

been presented by Denishev and Krumova (2005) and Mankame and Ananthasuresh (2001). In Chen et al. (2002), some simulation and experimental results are reported for this actuator type. Maximum deflection is observed as the cold-beam length reaches about 86% of the hot-beam length of a gold-plated actuator with dimensions of a 250x3.5x2μm hot beam and a 215x3.5x15μm cold beam. The deflection reaches 20.2μm at a driving current of 6.2mA.

An interesting thermal microactuator was recently presented in Ellis (2004). This actuator is designed as an element for a microassembly platform and it is actually a 3-D ECT-type actuator integrated in a single structure. This quadmorph microactuator incorporates four parallel cantilevers of equal length. Depending on the amount of heat the quadmorph can bend in plane or out of plane. The author claims that the platform with quadmorph actuators has four independent degrees of freedom, x, y, z and θ. There are no other details about the performance of this actuator, but it can be estimated that the overall parameters, such as speed and payload, are in the range of the previously described devices.

Shape memory

Shape memory alloys are metallic materials that demonstrate the ability to 'remember' and to return to their original, cold forged shape when subjected to the appropriate thermal procedure. Generally, these materials can be plastically deformed at some relatively low temperature, and upon exposure to some higher temperature, they return and hold on to their shape prior to the deformation. Thin film SMAs have been recognised as a new type of promising material for MEMS systems and especially for microactuators due to high power density, large displacement and actuation force and low operation voltage (Miyazaki et al., 2009).

Figure 2.5 shows a MEMS microwrapper manufactured from Ni-Ti (nickel – titanium) SMA (Gill et al., 2001). This device is not a cilia actuator, but with slight modifications, the microwrapper can be transformed into a 4–8 ciliary legs actuator. When current is applied via two opposite bonding pads, Joule heating flattens the legs, which are originally curled upwards in a cage-like formation. When the current is removed and the heat is dissipated to the environment, the legs return to their curled formation. The legs are 100μm long and 1.8μm thick and their displacement is significantly larger than the one of the previously described bimorph actuators. The price for the large displacement is large power consumption, though. A single actuator consumes 400mW, or, in other words, a matrix of 64 x 64 cells and with 10% of simultaneously active cells in the case of a massive processing will dissipate nearly 160W.

Figure 2.5 **Schematic of an SMA microrapper actuator in open and close position (adapted from Gill et al., 2001)**

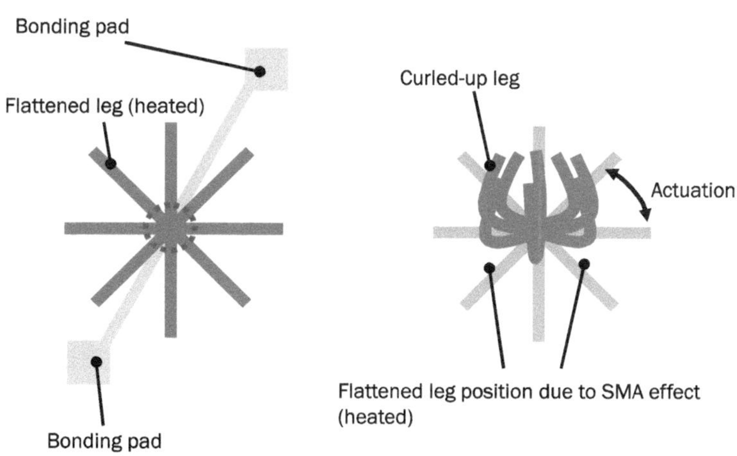

A different approach was presented in Yang and Gu (2008): two SMA wires are embedded in parallel with the axis of an elastic rod, whose one end is fixed on a substrate; this actuator is targeted for use in an anthropopathic robotic hand. The 'actuating' wire is set to obtain a 'U' memory shape when heated, whereas the 'restoring' wire is set to obtain a straight memory shape. By alternating heating between these two SMA wires, the rod can bend and return to its original vertical position, thus providing actuation similarly to the cilia actuators. The rod's radius is 5mm, the wire's radius is 250μm; the reported response time from maximum bending of the rod to its default state is 0.8s, so an estimation of a full actuation cycle time of 1.6s can be made. Three such rods – used as fingers of a robotic hand – can hold a sphere of 15mm radius and 3N weight and successfully manipulate it.

Cabrera et al. (2002) have presented a microactuator design of a thin film Ni-Ti cantilever with dimensions of 100x30x2.5μm. Simulations show that its tip can achieve a 39μm deflection with a thrusting force of 0.23mN and a cycle time of approximately 0.02s. The power required for the cantilever's actuation is 97.6mW. While this design was originally intended for the control of the opening and closure of a microvalve through the combined actuation rates of two cantilevers and a magnet, it could be used for ciliary planar micromanipulation.

Planar frictional

Linear frictional microconveyor driving units are equipped with three inverted feet, as shown in Figure 2.6 (Shay et al., 2008). Two feet ('Foot 1', 'Foot 3') are moved by thermal actuators of dimensions 400x10x2μm, whereas a third foot is fixed to the substrate ('Foot 2' – stationary). On top of the

Figure 2.6 Side view schematic of a 1-DOF frictional conveyor showing inverted foot motions, executing one step of the plate to the right (adapted from Shay et al., 2008)

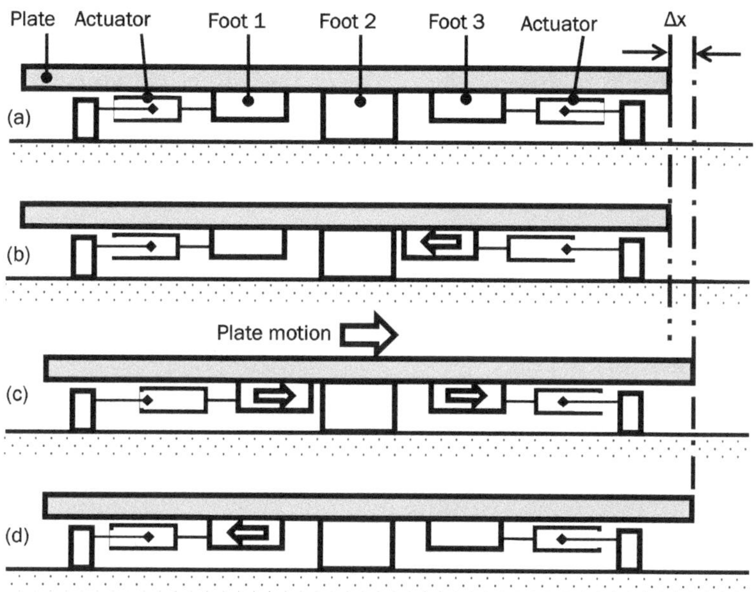

three feet rests a flat plate. By powering the thermal actuators that drive the feet, a friction force is created between the plate and the feet. When the feet are actuated in a suitable sequence, the plate will move in a stepwise manner. Figure 2.6 shows the motion phases required to carry out one step to the right. The plate can be moved in the opposite direction by reversing the order of steps. Additionally, a slightly different sequence of foot motions offers part rotation and two-dimensional translation on the workplane in two separate platform implementations. These conveyors are capable of reaching linear speeds as high as 33 and 20μm/s respectively and rotation rates as high as 7 degrees per second. Loads of 850mg on the plate at speeds of up to 13μm/s have been transported, with a displacement step of

2.5μm at 20Hz actuation frequency and with 15mW power consumption per actuator.

2.2.2 Electrostatic

Electrostatic charge arises from a build-up or deficit of free electrons in a material, which can exert an attractive force on oppositely charged objects, or a repulsive force on similarly charged objects. Electrostatic actuators can be of either contact or non-contact type, depending on the implementation. They are little affected by ambient temperatures and can generally exert large forces, but across very short distances.

Travelling electrostatic field

A multi-layered electrostatic film actuator was introduced in Egawa and Higuchi (1990) and Egawa et al. (1991). The device consists of a stator with electrodes and a slider with two layers: an insulating and a highly resistive layer. Figure 2.7 shows the principle of operation: initially, a charge distribution is induced on the resistive layer by applying a voltage pattern to the stator electrodes (a). The voltage pattern at the stator electrodes is then shifted to the right by one electrode (b). As a result, the attractive electrostatic forces between the charges on the stator elecrodes and the induced charges on the resistive layer shift the slider to the right, to the next equilibrium position (c). The realised motor by Niino et al. (1994) weighs 110g and consists of 40 stacked layers in order to maximise thrusting force. A maximum force of 8N and a power to weight ratio of 5W/kg have been achieved, using a driving voltage of 800V and 0.1Hz frequency. The outer dimensions of this multi-layered actuator are 20x14x50mm.

Figure 2.7 **Travelling field inductive surface drive motor (adapted from Egawa and Higuchi, 1990)**

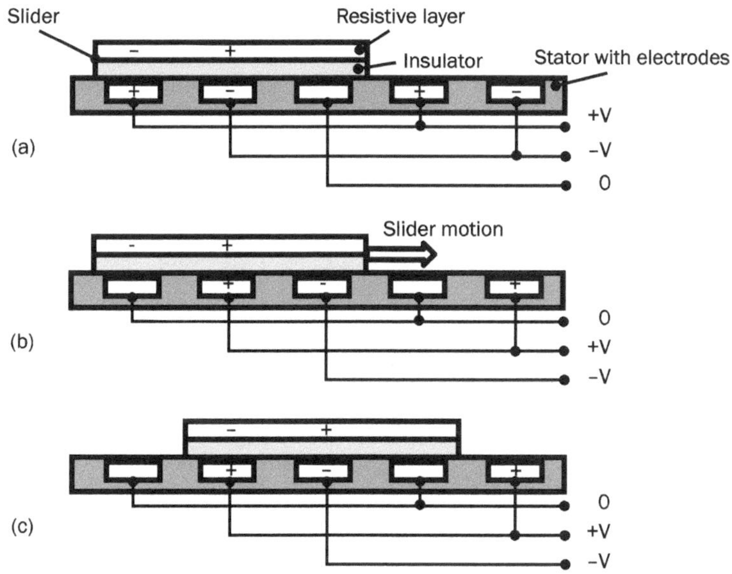

Comb-drive

A micro transportation system based on electrostatic comb-drive actuators and ratchet mechanisms was proposed in Dao et al. (2008). Various microcontainers of sizes of 450x250x30μm have driving-wings and anti-reverse-wings attached to the central 'backbone' and are driven by electrostatic actuators through the ratchet racks in perpendicular direction. The anti-reverse wings of the micro container act as ratchet teeth, in order to allow the container to move forward only. The experiments show that by applying a driving voltage of 100V with 20Hz frequency, the containers' maximum velocity reaches 200μm/sec with a displacement per ratchet tooth of approximately 15μm.

The working principle can be seen in Figure 2.8. In (a), where the straight module is presented, the left and right

Figure 2.8 Configurations of (a) the straight module, (b) the turning module and (c) the separation module (Dao et al., 2008; © [2008] IEEE)

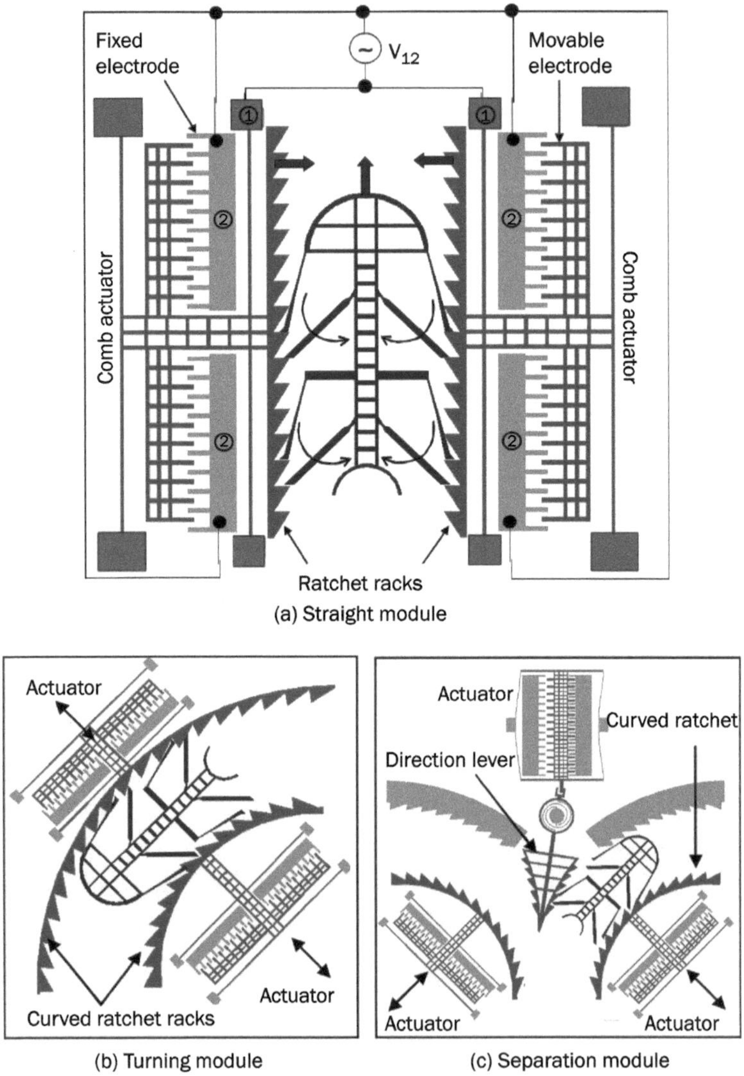

ratchet racks are connected to the movable electrodes. When a periodic voltage between the movable and fixed electrodes is applied, the left and right ratchet racks will move back and forth with the same frequency and displacement but in opposite directions. As a result, the container moves forward in a perpendicular direction. When the driving voltage becomes zero, the micro container does not move backwards because of the anti-reverse wings, which always engage with the ratchet teeth. After each cycle, the container moves at least one pitch of the ratchet teeth. The procedure for the turning (b) and separation (c) modules is similar.

Scratch-drive

This actuation principle was introduced by Akiyama and Shono (1993a, 1993b) and is illustrated in Figure 2.9. It consists of a thin (1–2μm) polysilicon plate, with a so-called

Figure 2.9 **Actuation principle of the scratch drive actuator (adapted from Akiyama and Shono, 1993a)**

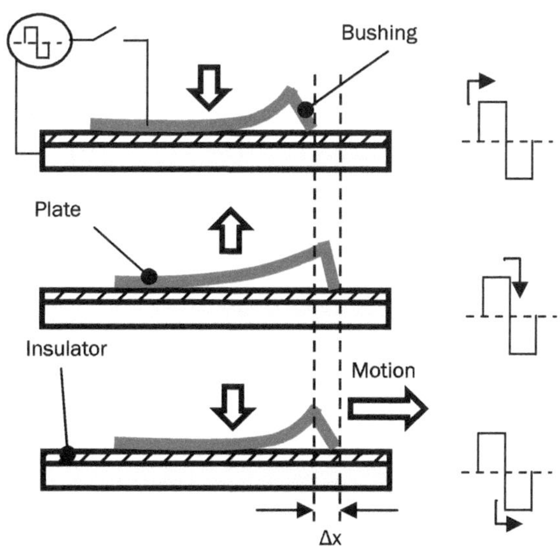

'bushing' at one end of the plate. In the activated state the plate is attracted to the electrode of the substrate. As the actuation voltage switches from positive to negative, the plate relaxes shortly and the bushing rotates. Because the friction between the bushing and the insulator film is higher than the one between the plate and the substrate, the bushing does not slide and the centre of rotation is the line where the bushing touches the insulator. As a result, the plate shifts slightly forward. The actuation cycle can then be repeated anew. Bronson et al. (2004) conducted experiments with scratch actuators of various dimensions, with the most notable results being an average step of approximately 31μm for a 200x65μm plate and an electrostatic force of 117μN for 200V applied voltage with 100Hz frequency.

Electrowetting-on-dielectric

An alternative proposal for microconveyance based on electrowetting-on-dielectric (EWOD) was introduced in Moon and Kim (2006). Using the movement of liquid droplets by EWOD actuation instead of active mechanically moving components, an effective mechanism of a microconveyor system was realised. Figure 2.10 shows a schematic of the actuation concept. The actuated droplets serve as the moving wheels of a conveyor pad. The movement of the droplets is controlled by a simultaneous voltage application on the electrodes, which are next to each droplet. The induced electric field modifies the wetting behaviour of the droplets in contact with the electrodes and as a result they are stretched and then pulled onto the next electrodes. The conveyor pad is carried along with the moving droplets.

The actuation voltages are in the range of 80V_{AC} (at 1kHz square wave) for 5.5mm×5.5mm×40μm comb-shaped electrodes, while the reported unoptimised transportation

Figure 2.10 **Actuation principle based on EWOD (adapted from Moon and Kim, 2006)**

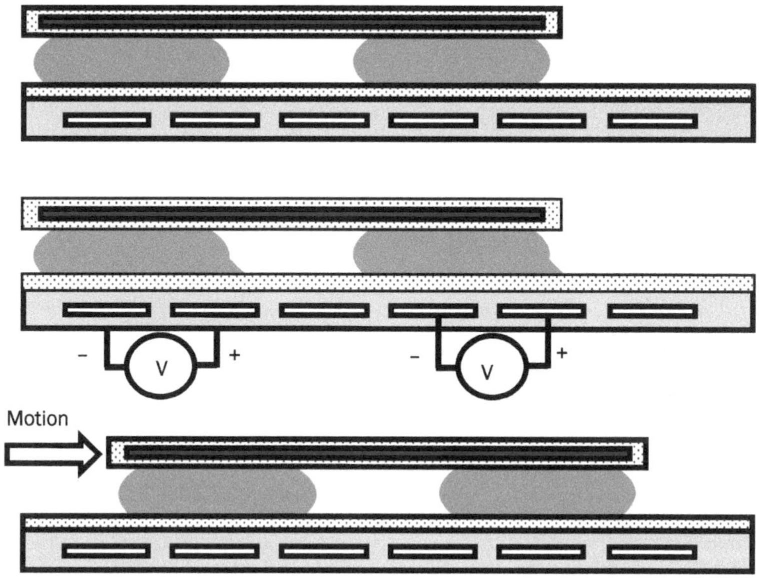

speed reaches 2.5mm/s. Transportation of a variety of objects has been tested, including other water droplets on the conveyor pad and several silicon chips of total weight of approximately 180mg. The main advantages of this implementation are: the possibility for fast transportation speeds as well as flexibility of handling microparts of various shapes, geometries and dimensions since manipulation occurs on the droplets and not on the parts themselves.

2.2.3 Electromechanical

The main types of electromechanical actuators are electromagnetic and piezoelectric. Electromagnetism arises from electric current moving through a conducting material. Attractive or repulsive forces are generated adjacent to the conductor and proportional to the current flow. These actuators harness electromagnetic forces in order to create motion, often with large deflection capabilities. Piezoelectric motion arises from the dimensional changes generated in certain crystalline materials when subjected to an electric field or to an electric charge. Such actuators respond very quickly to changes in voltages and with great repeatability, usually have low power consumption and can be driven at high speeds, but generally they also exert low forces and small strokes.

Electromagnetic

In Liua et al. (1999) millimetre-scaled magnetic actuators have been developed, capable of achieving large out-of-plane angular displacements and forces. Each actuator consists of a 1x1x0.005mm Permalloy ($Ni_{80}Fe_{20}$) piece attached to two 400x100x1μm cantilever beams whose other end is fixed on the substrate. Under the influence of a magnetic field created by an electromagnet, this actuator – originally resting horizontally on the plane – can achieve 65-degree angular and 1–2mm vertical displacements and exert an 87mN force in the direction perpendicular to the substrate. Although this actuator primarily targets fluid mechanics applications (actuator flaps for fluid dynamic control), it could be potentially used for microconveyance as a contact actuator, similarly to the cilia ones.

In a similar implementation, an electromagnetic actuator was introduced in Judy and Muller (1996, 1997). This actuator consists of a torsional polysilicon beam, firmly fixed at both ends to the substrate, and an attached magnetic plate made of NiFe at the middle of the beam. When a magnetic field is applied by an electromagnet, a pure moment is induced and the magnetic plate is rotated out of plane (Figure 2.11). The electromagnet is actually a 10-turn coil

Figure 2.11 **Operation principle of the electromagnetic actuator (adapted from Moon and Kim, 2006)**

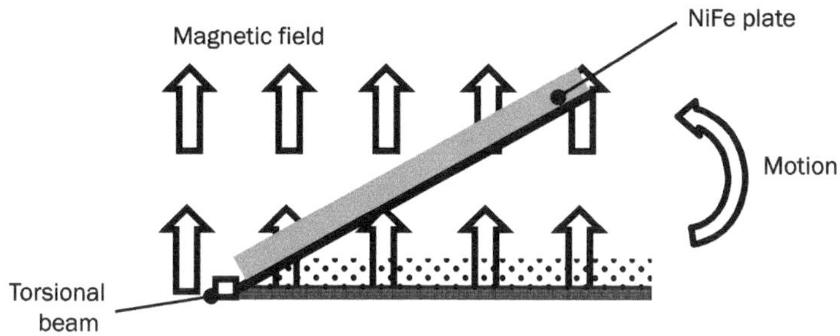

integrated on the substrate and surrounds the moving structure. The sizes of the actuator's plate and torsional beam are 430x130x15μm and 400x2.2x2.2μm respectively, whereas the entire device (actuator and coil) requires an area of 1.4x1.4mm. The measured deflections reached up to 90 degrees with a torque of 3nN-m for a magnetic field of approximately 10kA/m but the power needed is high (about 6.25W). For microconveyance, this actuator seems limited by the tolerance of the junction of plate and beam under external load (micropart) as well as the high power requirements.

Piezoelectric

A piezoelectric ultrasonic bidirectional linear actuator for micropositioning was presented in Friend et al. (2008) which comprises a piezoelectric block and a pair of vibrating slanted beams with circular shapes at their tips (Figure 2.12). One of these shapes is slightly larger in diameter, causing the resonance frequencies between the two beams to be different. The beams' flexural vibration combined with the vertical vibration from the piezoelectric element gives elliptical vibration of one of the beam tips, depending on the driving frequency and the relative phase of the vertical and flexural

Figure 2.12 **Schematic of the linear piezoelectric ultrasonic actuator (adapted from Friend et al., 2008)**

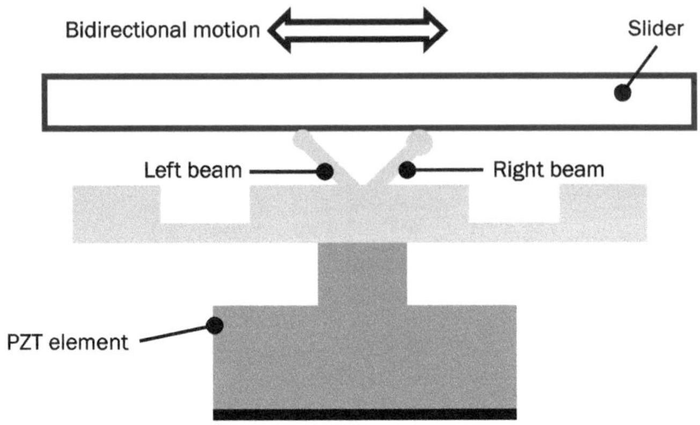

vibrations. By switching between the resonance frequencies of the two beams, the direction of sliding may be reversed. This actuator offers bidirectional motion at an excitation frequency of about 1.69MHz and 1.71MHz for the two beams and 1.7MHz for the piezoelectric element. The forces generated at the right and at the left beam reach 80mN and 42mN respectively, at speeds of up to 40mm/s using a large 23g polished alumina slider.

Another interesting type is the bending mode actuator, also referred to as 'wobble motor' (Watson et al., 2009; Kanda et al., 2004; Koc et al., 2002), since due to resonant bending its tip executes an elliptical motion while its base remains firm on the substrate. Such actuators are usually shaped as beams (resembling pillars if they are vertically placed on the substrate). The way these beams are bent can be isolated by driving the piezoelectric element at suitable frequencies; this could be a promising aspect since motion towards the x or y direction exclusively can be achieved (Figure 2.13). Additionally, by allowing the full elliptical patterned motion of an array of these actuators, part rotation could be also achievable. A characteristic example is the

Figure 2.13 Bending modes of the 'wobble motor' type actuators (adapted from Watson et al., 2009)

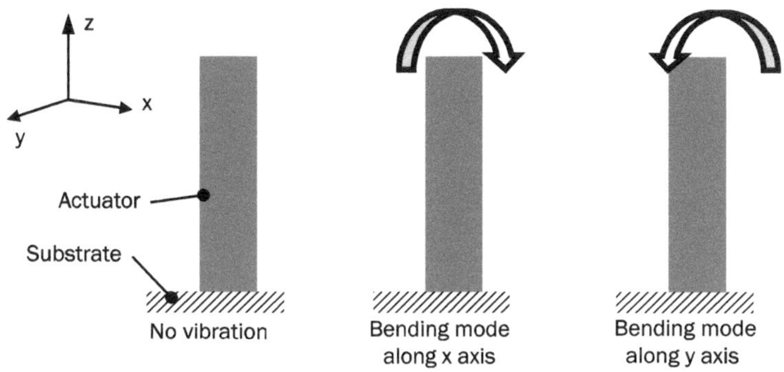

actuator introduced in Morita et al. (1999), which produces a maximum rotational velocity of 650rpm and a maximum torque of 220μNm at 85KHz driving frequency and $100V_{p\text{-}p}$ input voltage.

2.2.4 *Fluidic-mechanical*

This type of actuator platform relies on contactless pneumatic actuation, where the objects are kept away from the actuator surface by means of airflow. This allows simultaneous levitation and conveyance of parts. Generally, these platforms offer high velocities and absence of friction problems, but in return, they require a greater level of complexity control. The main problems are flow instabilities and turbulences, which necessitate the use of sensors and imaging in order to minimise their effect.

Pneumatic

Pneumatic microconveyor platforms are contact-free systems since pneumatic force fields are used to create a cushion to separate the object from the surface. Figure 2.14 illustrates

Figure 2.14 **Principle of airflow manipulation (adapted from Chapuis et al., 2005)**

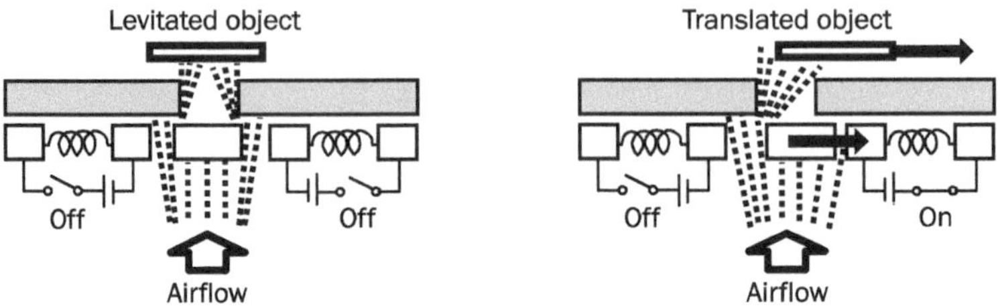

the principle of airflow manipulation, adapted from Chapuis et al. (2004). Although the microactuator cell in this system has one degree of freedom, cross-arranged cells within the array enable two degrees of freedom movement of objects. The top substrate has a through-hole for airflow, while the bottom substrate has a moveable part with suspension beam electrodes that works as a normally closed valve.

A two-dimensional pneumatic microconveyor platform was presented by Zeggari et al. (2010). The conveyor uses inclined air jets that can levitate and move flat objects to a desired position using controlled airflow. The airflow comes out of planar, square-shaped holes which contain four 250x60μm micronozzles each, thus allowing four different conveyance directions (east, west, north, south). Nozzles corresponding to a conveyance direction are linked together and blow simultaneously. For the experiments, a 3mm diameter, 0.5mm thick silicon chip weighing 2mg is placed on the platform. The silicon chip starts to move when the air pressure reaches 20kPa and the mean velocity of conveyance reached the value of 2–3mm/s. Figure 2.15 shows the steps of a manipulation procedure.

A similar implementation was introduced in Delettre et al. (2010). This platform is able to move an object on an air-hockey table based on a novel traction principle. The object is pulled in and moved indirectly by an airflow which is

Figure 2.15 Conveyance of a silicon chip using periodic compressed air. The blowing direction is represented by an arrow on the object (Zeggari et al., 2006)

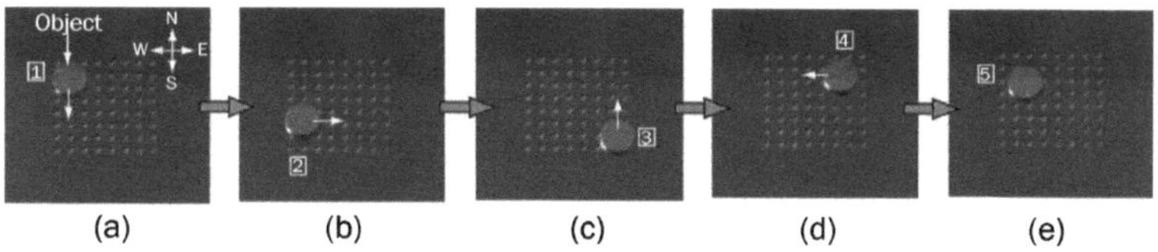

induced by strong vertical air jets coming from specific 400μm diameter nozzles of the air-hockey table. The operating pressures are 10kPa for levitation and 500kPa for traction and the part used for one-dimensional conveyance is an aluminium cylindrical object of 30mm in diameter, 15mm height, with a mass of 29g. The maximum observed velocity of the part reached approximately the value of 100mm/s.

An alternative concept platform called 'active surface' was presented by Ku et al. (2001), which integrates automation concepts in handling multiple objects with high precision. The manipulation principle relies on the existence of two types of pressure: positive (blowing) pressure and negative (vacuum) pressure. The blowing pressure creates an ascending force to decrease the friction between object and active surface by lifting the object slightly above the surface. Conversely, the vacuum pressure creates a descending force, which draws the object to the surface and maintains it in place. The presence of one blowing tube next to a vacuum tube forms an airflow which exerts a force directed from the high-pressure area to the low-pressure area. If this force is larger than the friction force, the object will move in the direction of the airflow force. In this implementation, an object recognition algorithm and several control techniques are used to perform basic sensing and handling operations of parts, such as translation and rotation, but no further details are provided.

2.2.5 Comparison and suitability for planar manipulation

Various implementations of actuators and actuator microconveyor platforms that could potentially be used for planar manipulation of microparts have been presented so far. Table 2.1 shows a comparison of the most indicative of these implementations, along with all the reported information regarding their design and performance characteristics, i.e. actuator dimensions, maximum velocity achieved, force or torque per actuator or pressure, the load of the object used for conveyance experiments, the horizontal displacement per actuator as well as the actuation frequency, the power consumption per actuator and finally an estimation of the fabrication and the operation complexity.

An analysis of the suitability of the various approaches can be made using the criteria mentioned at the beginning of this section. All of the characteristics of Table 2.1 are important, but the main focus can be given in large actuator strokes (large displacements) and large forces, without however neglecting the rest. The ciliary thermal bimorph and polyimide v groove thermomechanical actuators provide a very good correlation of propulsion force, displacements, speed and load capabilities. Unfortunately, not enough details are available for the performance of the ECT actuator and the elastic bending rod with SMA wires for comparison. Planar frictional platforms provide small displacements with good actuation frequencies and therefore relatively fast conveyance speed; additionally, they can perform not only two-dimensional translation but rotation as well. The common issue that all the thermal actuators face is the heat dissipation after their activation. Dissipation into the environment is a simple and common practice, but in turn limits the actuation frequencies since it delays the activation/deactivation cycle.

Table 2.1 Comparison of the most indicative implementations of actuators and actuator platforms for microconveyance

	Authors	Description	Actuator dimensions	Maximum velocity (mm/s)	Force or torque per actuator or pressure	Moved object or load	Horizontal displacement or motion step and actuation frequency	Power consumption per actuator	Estimated fabrication and operation complexity
Thermomechanical	Suh et al., 1997	Array of erected triangle-shaped thermal bimorph actuators	437μm length, 105μm width at base	1	23μN	3x3x0.1mm silicon die of 1.7mg	95μm, 0.5Hz	35mW	High, high
	Wu and Hsu, 2006	Array of erected thermal bimorph actuators	400x50x4.5μm	1.5	**	0.196mg per actuator	18μm, 5Hz	**	Medium, low
	Ebefors et al., 2000	Array of erected legs-thermal actuation of polyimide joints	500x600x30μm	12	100mN	350mg per actuator, 14x7x0.5mm Si pieces and external load of 2,100mg	170μm, 3Hz	175mW	Medium, low
	Chen et al., 2002	ECT actuators	250x3.5x2μm hot beam and 215x3.5x15μm cold beam	**	**	*	20.2μm, **	**	Medium, low
	Yang and Gu, 2008	Elastic rod actuator with embedded SMA wires	5mm rod with 250μm SMA wires	**	**	15mm radius sphere of 3N weight for 3 actuators	**, 0.625Hz	**	Medium, low

(*Continued*)

Table 2.1 *(Continued)* **Comparison of the most indicative implementations of actuators and actuator platforms for microconveyance**

	Authors	Description	Actuator dimensions	Maximum velocity (mm/s)	Force or torque per actuator or pressure	Moved object or load	Horizontal displacement or motion step and actuation frequency	Power consump-tion per actuator	Estimated fabrication and operation complexity
	Shay et al., 2008	Planar frictional platform	400x10x2μm	13 (up to 33 with no load at 500Hz)	**	580x960μm plate plus cylindrical graphite of 850μg load	2.5μm, 20Hz	15mW	Medium, medium
Electrostatic	Niino et al., 1994	Multiple layer travelling electrostatic field actuator	20x14x50mm	**	8N	**	**, 0.1Hz	power to weight ratio of 5W/kg	Medium, medium
	Dao et al., 2008	Comb-drive actuator and ratchet mechanism platform	134 fingers of 40x3μm, 700x4.5μm beams, 6x10μm ratchet teeth	0.2	**	450x250x30μm microcontainers	15μm, 20Hz	**	High, high
	Bronson et al., 2004	Scratch-drive actuators	200x65μm	*	117μN	*	31μm, 100Hz	**	Medium, high
	Moon and Kim, 2006	EWOD electrode platform	5.5×5.5×0.04mm comb-shaped electrodes, 6μl droplets	2.5 unoptimised	**	Up to 180mg silicon chips stack	Relative to applied voltage control and electrode size	**	Medium, medium

Electromechanical	Liua et al., 1999	Electomagnetic actuators	400x100x1μm beams with 1x1x0.005mm Permalloy piece	**	**	*	65-degree angular, **	**	High, medium
	Judy and Muller, 1997	Electomagnetic actuators	430x130x15μm magnetic plate, 400x2.2x2.2μm torsional beam	**	3nN-m	*	90-degree angular, **	6.25W	High, medium
	Friend et al., 2008	Bidirectional piezoelectric linear actuator	Less than 400μm^3 stator	40	80mN rightward, 42mN leftwards	0.57x2.07x50mm aluminium slider of 23g mass	**,*	**	Medium, low
Fluidic-mechanical	Zeggari et al., 2006	Array of inclined air jets	250x60μm nozzles	3	20kPa	2mg silicon chip of 3mm diameter, 0.5mm thickness	**, 1Hz	**	High, high
	Delettre et al., 2010	Array of air jets for levitation and traction	400μm diameter nozzles	100	10kPa levitation, 500kPa traction	29g aluminium cylinder of 30mm diameter, 15mm thickness	*,*	**	High, high

* not applicable, ** not reported

Reported details for the presented electrostatic actuators are not enough in order to be able to actively compare them to the other types. Besides, due to design, there are limitations on their practical application for planar conveyance of microparts: the scratch drive actuators are not directly applicable for part conveyance and would probably need a new design specific for planar conveyance, whereas the comb-drive ones with ratchet mechanism essentially limit the motion of microcontainers on 'rails' only. The travelling electrostatic field platform is an interesting concept offering large forces (and presumably actuation speed), but with apparently large power consumption (as evidenced by the 800V applied voltage) and a limitation of proper material selection for the conveyed parts. EWOD actuation with a carrier pad has the potential of very fast two-dimensional conveyance speeds but the main limitations are that only conveyance is feasible (no part rotation) and that a microgripper is needed for loading and unloading of parts.

Electromagnetic actuators could provide contact actuation on microparts (similar to thermomechanical actuators), with large angular, out-of-plane deflections. However, there could be problems on an array placement, as the magnetic field of one activated actuator could also affect and activate neighbouring ones for cases of asynchronous driving. There should not be such problems for synchronous driving where all actuators are simultaneously activated/deactivated, though such an option limits manipulation flexibility as well as speed. Additionally, the issue of load tolerance in the bending areas of the actuators should be taken into consideration, as well as the power required for the electromagnets. The presented piezoelectric actuator on the other hand can offer fast manipulation speeds and large forces for part conveyance along one dimension. Realising conveyance in two dimensions can prove difficult, however, and most probably a redesign would be needed. The piezoelectric 'wobble motors' seem to have potential but they need to be tested with the concept of part conveyance in mind.

Fluidic-mechanical platforms consist of variations of air jet arrays and can offer fast to very fast speeds with no friction and very good load capacities but are mainly limited by the complexity in fabrication and operation as well as the necessity for control of the part's motion. Constant airflow can generate strong instabilities due to fluid turbulences. Fluid effects are also very non-linear and it is often hard to model precisely interactions between tens of different air jets. Applying coherent control laws for hundreds of independent air jets is thus a complex problem.

We can therefore conclude that currently the most suitable types of actuators for micromanipulation on a plane are the thermomechanical actuators, followed by the fluidic-mechanical actuators. In particular, the big advantage that the bimorph and polyimide actuators (and in extent ciliary SMA actuators such as the microwrapper) have is that they can be individually addressed through proper control and programming circuits. A control system for selective activation of actuators on a platform can provide specific actuation gaits (asynchronous driving) which are faster than normal conveyance gaits (synchronous driving), as well as advanced operations such as sorting, alignment and rotation of parts. Although a control system requires more complex platform design, the flexibility it can offer is enormous. Additionally, use of SMA ciliary actuators for larger deflections than normal bimorph materials (Ren et al., 2000) as well as a proper cooling system (such as heat dissipation fins on the ciliary legs or air cooling) for increased actuation rates can significantly improve performance.

2.3 Manipulation of parts on microconveyor platforms: an integrated approach for programmable force field design and platform programming

One of the most promising alternatives for manipulation, sorting and conveyance operations on actuator microconveyor platforms has been the introduction of programmable force fields or, in other words, mathematical expressions of force fields produced by functions of potential or by other methods. They are characterised as 'programmable' because most of them can be diffused/applied on a microconveyor platform by proper programming of the actuator cells, so that they induce motion as directed by the diffused field. When a part is placed on a programmed platform, it is translated and rotated until it reaches a stable equilibrium pose (position and orientation) due to the force and torque applied by the actuators. This approach is very favourable for the sensorless manipulation of a wide variety of microparts since little or no sensing is required in order to handle microparts towards assembly automation (Böhringer et al., 1996; Luo and Kavraki, 2000).

Many types of programmable force fields have been proposed, ranging from relatively simple ones to more complicated ones. Luo and Kavraki have shown how to perform assembly of two parts using radial, curl and push fields, offering centring, rotation and translation respectively, as seen in Figure 2.16 (Luo and Kavraki, 2000). Böhringer and Donald showed that the successive application of a predetermined number of squeeze fields – depending on the part complexity – can translate and re-orientate a micropart towards a stable equilibrium position (Böhringer and Donald, 1996). In Sudsang and Kavraki (2001), the authors

Figure 2.16 Phases of an assembly procedure: (a) initial positions and orientations of two parts, (b) centred parts with radial fields, (c) application of curl fields, (d) desired part orientations after rotation, (e) push field applied to one part, (f) assembled parts (adapted from Luo and Kavraki, 2000)

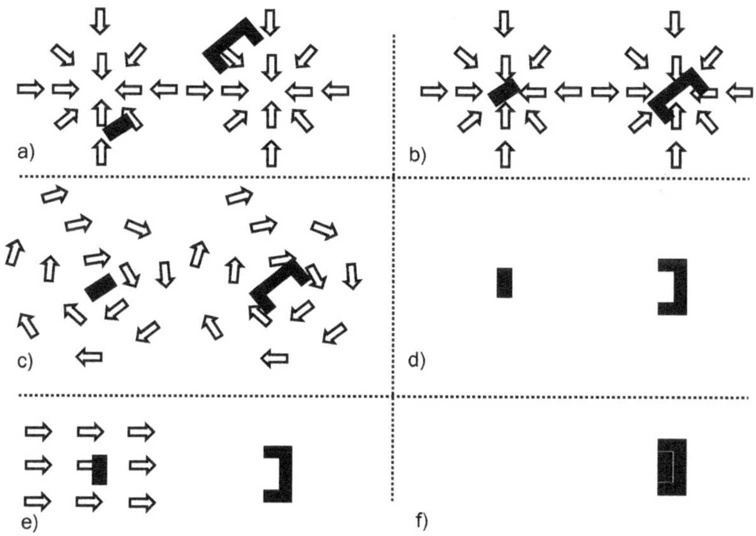

proposed a combination of a linear radial field and a constant field that induces a unique stable equilibrium for any non-symmetric micropart. In Lamiraux and Kavraki (2001), the combination of a unit radial field with a small constant field was introduced for the positioning of symmetric and non-symmetric microparts along with the computation of all possible equilibrium configurations. A two-dimensional field was presented in Xidias and Aspragathos (2006a), which induces a stable equilibrium for any convex or non-convex polygonal micropart from a polynomial potential function, considering that the desired final position and orientation of the micropart is the unique stable equilibrium. In the same sense, a field design method based on solid geometry concepts was introduced in Xidias and Aspragathos (2006b): in the region inside the desired location of the micropart the force applied is zero, while outside the vector field is other than zero and directed towards the desired location.

A new method was introduced for the automatic formulation of a programmable force field that can be used for the manipulation of microparts (Lazarou and Aspragathos, 2009). Under this framework, microparts of various polygonal shapes can be positioned and orientated on the same platform in a sensorless, open-loop procedure without changing the hardware or software. The concept as well as simulation results are presented below.

Force field generation

In order to design the force field, only the knowledge of the part's geometry and its desired final position and orientation is required. The initial position and orientation of the part can be unknown and anywhere inside the field area since the applied force and torque move it to the desired pose.

Assume the final pose of an N polygonal part is bounded by N straight line segments, each corresponding to one edge of the polygon. Each such line divides the plane into two half-planes and the final pose is defined as the intersection of all the half-planes lying to the right of the polygon's edges, if its boundary is followed in a clockwise direction. For all the points of each line and its left half-plane the local force field is constant and with direction normal to the line, whereas in the right half-plane it is zero (Figure 2.17).

The final force field is constructed as the vector aggregation throughout the whole plane and it essentially 'squeezes' the whole final pose compared to the squeezing field along a line by Böhringer and Donald (1996). The part is pushed and rotated towards this final pose until it is 'trapped'/'locked' in

Figure 2.17 Final desirable pose of an N-polygon, bounding lines and an indicative field of one of them

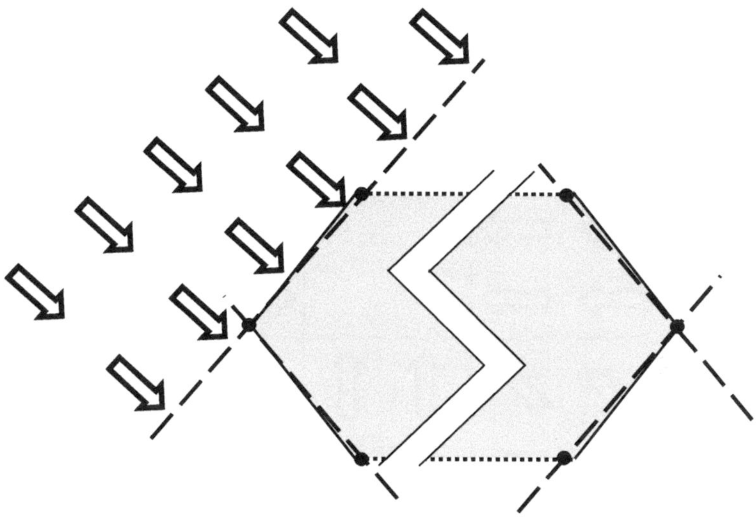

the target position with the desired orientation. Therefore, the final pose is an equilibrium pose of the particular micropart. If the micropart is symmetrical, there can be multiple equilibrium poses (orientations) located in the desired position whose number depends on the part's geometry, as seen in Figure 2.18. The point A can be in any of the four corners so the possible different orientations are four. At the moment there is no mechanism to drive the micropart to a particular desired orientation. In the cases of non-convex polygons, they are first decomposed into convex polygons; the procedure is the same thereafter and the final field is produced again by the vector aggregation.

Field diffusion, actuation principles and programming of microconveyor platforms

It is assumed that square 'motion pixels' are uniformly spread on the workplane (the surface area of the microconveyor

Figure 2.18 Stable equilibrium pose of a square micropart, corresponding half-planes and the final force field

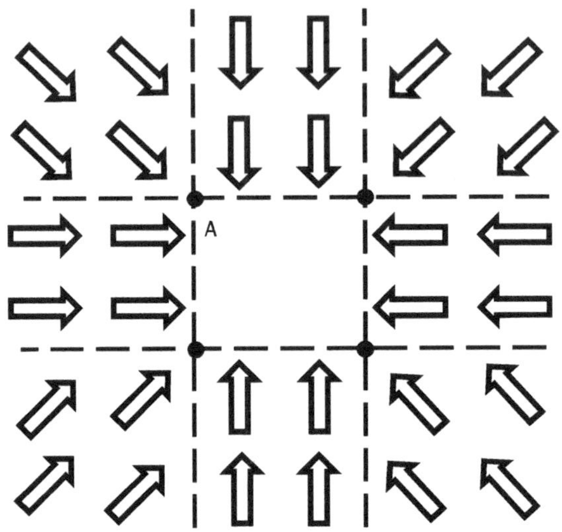

array) and that every pixel includes four orthogonally placed actuators, similar to the cilia bimorph platforms. Each motion pixel is assigned with a specific force field vector and therefore induces motion towards that specific direction. In the cases where the vectors form arbitrary direction angles with the x-axis, the shape of the final desired pose of the part is approximated with an accuracy depending on the selected pixel size.

After the diffusion of the field on the array, if a motion pixel is assigned with a vector pointing in one of the four prime directions, the actuation easily takes place by activating the corresponding actuator of the pixel. In cases of arbitrary vector angles with the x-axis the necessary equivalent ratio of motion along the x and y directions is considered (for example 45-degree angle requires motion ratio of 1/1). A good approximation of arbitrary angles can be achieved by using the integers 1, 2, 3 and 4 and by therefore having four different actuation rates in the x and y directions respectively or, alternatively, how many times the actuators are activated per one period of the actuation pulse signal. The user has to consider the hardware complexity when defining the maximum error and thus a compromise is necessary.

The programming of the platform is performed serially, through appropriate digital circuitry (Lazarou and Aspragathos, 2009). For each motion pixel, the angle of the diffused force field vector dictates the approximate motion ratio in the x and y axes; the quadrant of the angle also indicates which of the four actuators of the motion pixel should be activated. In the motion pixels inside the final pose all actuators are inactive. When the programming of all motion pixels is finished, the platform starts its operation for the actual part manipulation.

In each motion pixel outside the final pose, all the actuators are activated periodically, in order to apply force and induce motion on the micropart according to the diffused field. The activation of all these pixels is parallel/simultaneous. A simple two-phase gait is selected for the activation of the specific actuators according to the field formulation for the particular micropart. The two states ('on' and 'off') are continuously alternated and move the micropart that lies atop them with small 'strokes' from its initial pose towards the desired pose.

A brief example is presented below. A trapezoid-like micropart and an 8x8 platform are selected for the programming. The diffused force field vectors and the desired and actual final pose are shown in Figure 2.19. The field is divided in eight regions of different vector directions. In region 5, the field's vector angle is 116.25°, or a motion ratio of 2/1. Thus the needed effect is two strokes from the bottom actuator (pushes upwards) and one from the right (pushes leftwards). In a similar way, the actuation ratios in regions 6 and 7 are respectively for the bottom and right actuators 3/4 and the top and right 1/2.

Figure 2.19 Top: Use of a trapezoid-like micropart for the programming of an 8x8 array; Bottom: region 5 motion pixel actuation rates (Lazarou and Aspragathos, 2009)

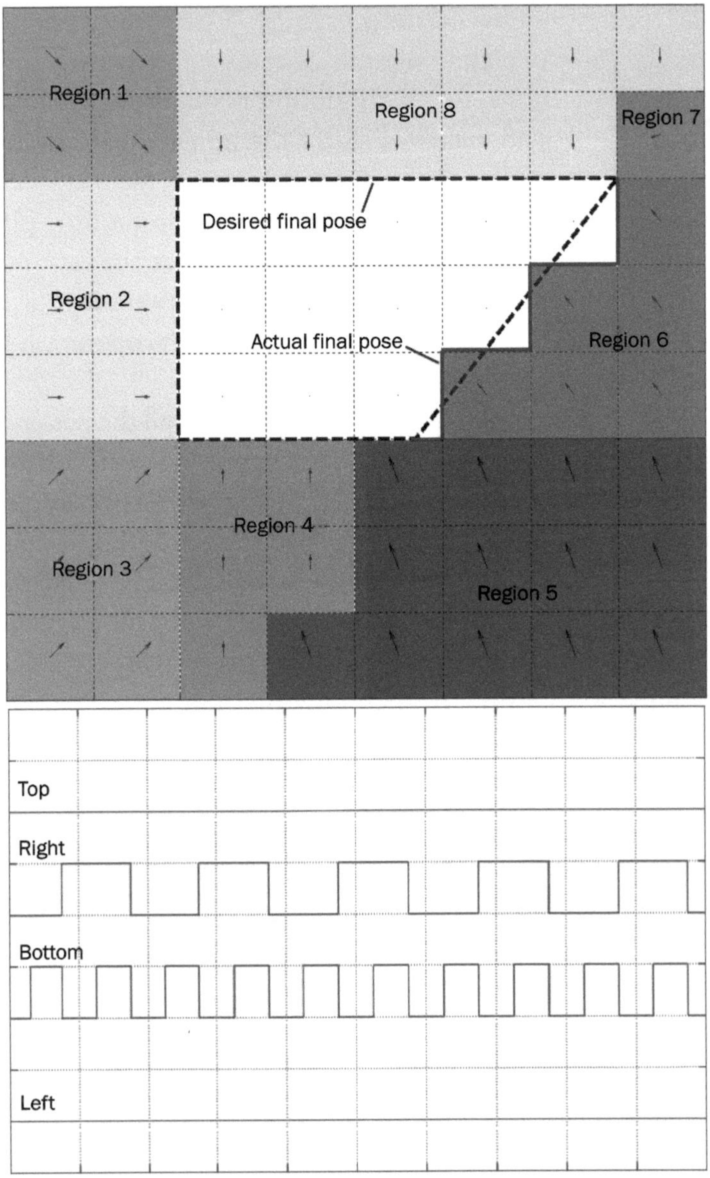

Simulation for polygonal parts

An application has been developed that simulates the motion of a part on a microconveyor platform under the derived planar force field. The platform's dimensions are 4.2x4.2mm, containing 21x21 motion pixels of dimensions 0.2x0.2mm each. The period of actuation according to Böhringer et al. (1997) is 2s, which corresponds to four actuation rates of 2s, 4s, 8s with three strokes and 8s with four strokes.

An example of a non-convex triangular-like micropart is shown in Figure 2.20, with length of edges

Figure 2.20 Final desired pose and corresponding force field (Lazarou and Aspragathos, 2009)

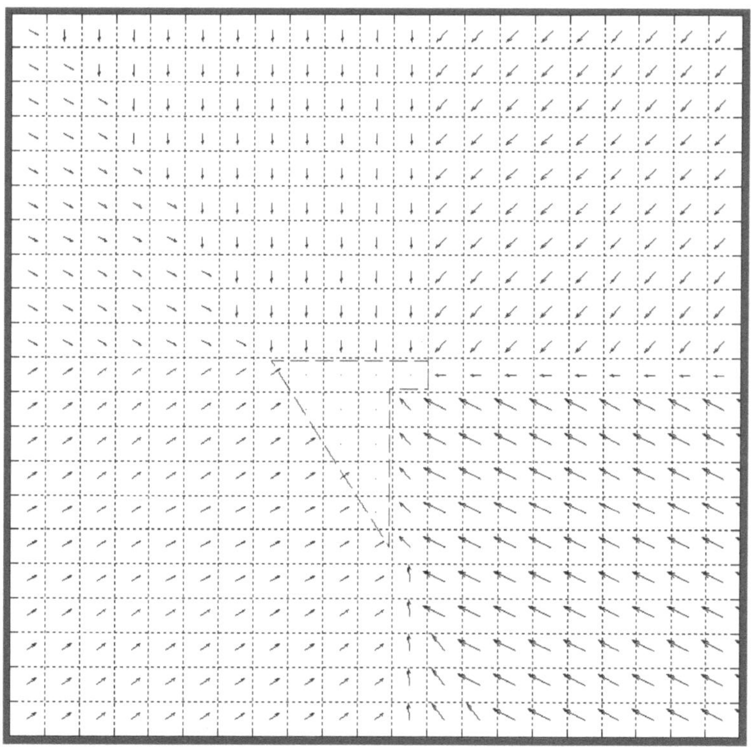

Figure 2.21 Part's position and orientation over time (Lazarou and Aspragathos, 2009)

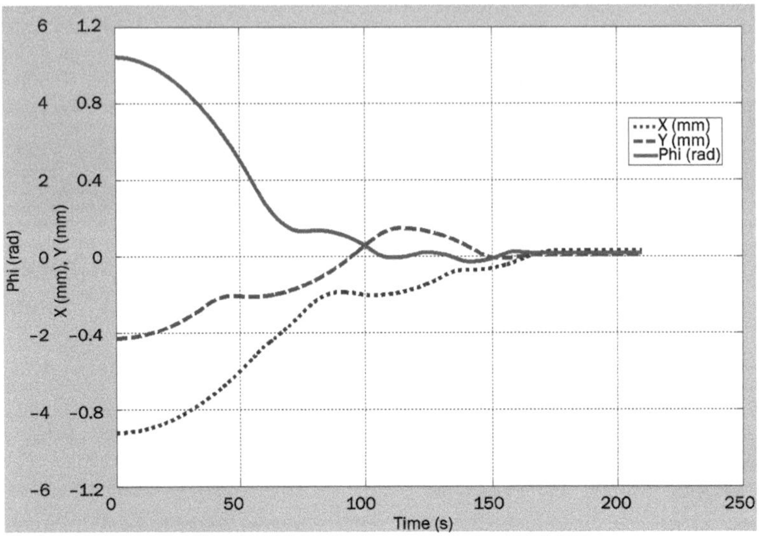

0.9x0.18x0.21x0.91x1.29mm. For the initial location (0.92, 0.428) (mm) and angle of 5.2 rad (298°), the part reaches the point (0.028, 0.008) (mm) with an angle of 0.089 rad (5°) after approximately 162s, as seen in Figure 2.21. The positioning error is 28μm in the x direction and 8μm in the y direction and the orientation error is 5°.

Advantages and limitations of the proposed approach

With the introduced method for field design a field can be derived for manipulating a micropart of any polygonal shape. In the simulations, the part reaches the desired location and orientation with relatively small deviations and within the pixel resolution. If the microparts are asymmetrical or non-convex, they rest in one unique positioning and orientation equilibrium. The manipulation time is relatively long depending on the reported actual cilia actuators' thermal expansion/contraction low rates. Consideration of different types of contact actuators, such as shape-memory or piezoelectric, would drastically improve the actuation frequencies and thus make the proposed approach much more appealing for micromanipulation.

The advantages of the proposed integrated approach can be summarised in the following:

- sensorless, open-loop manipulation, as the micropart motion is dictated purely by the programmed field on the platform;
- reduced cost and complexity, especially in the case of sorting or manipulating many microparts in parallel on an array;
- programmability, repeatability and flexibility; the process is completely programmable and manipulation of microparts of different geometries is feasible on the same platform without changing the hardware, by simply feeding the actuators with new data, derived by the generated field according to the new micropart's shape.

2.4 Future research directions and conclusions

A variety of microconveyor platforms based on microactuators has been presented in the last two decades. The reported conveyance velocities range from μm/s up to around a hundred mm/s and are a correlation of the respective actuator deflection (stroke) and its actuation frequency. In general, large deflections with at least a frequency of a few Hz tend to give faster manipulation speeds. In most of the presented cases, the reported velocities can be deemed quite promising.

While actuator deflections and manipulation speeds are generally on a good track, there are still some important issues that have to be addressed, such as actuation cycle and payload capacity, since they are often limiting factors. Therefore, the emphasis of future research should not only be placed on the optimisation of the existing actuator characteristics but also on the creation of new, improved designs from possibly new materials. Especially for thermomechanical actuators, such as bimorph, polyimide V-groove joint and SMA, actuation rates can be definitely improved with better design for passive heat dissipation into the environment (for instance with the addition of fins that increase their overall surface area) or with cooling systems (air, water, etc.).

In addition, more advanced techniques and strategies should be developed in order to successfully control the manipulation of microparts on a platform. One of the main objectives should be the sophisticated manipulation (translation, rotation, sorting) of multiple parts on the same platform without or with little sensing in order to keep both the cost and the complexity at low levels. Towards this end, the programmable force fields can offer a solution but with certain compromises. The main problem lies in the diffusion of the field onto the platform's actuator cells, or in other words in the cell dimensions (smaller cells and actuators provide better, finer motion but their load capacity is smaller). Therefore, there are limitations set by the hardware components of each platform. However, if a platform is designed on hardware level (dimensions, payload, arbitrary motion directions) with the target of using programmable fields, the potential of parallel manipulation of many microparts of various geometries could prove very promising.

This chapter has provided an overview of the recent developments for manipulation of microparts on microconveyor arrays. While microcomponents are nowadays fabricated in batch quantities, their sorting, transportation and alignment procedures for the assembly of the final MEMS devices are still lacking in efficiency. This is an area where microconveyor platforms could offer a solution, as they seem promising for parallel manipulation if design and actuation issues are addressed. The various platform actuator types are classified based on their input energy: thermomechanical, electrostatic, electromechanical and fluidic-mechanical. For each such class, a variety of existing platforms or promising implementations are presented and a comparison is made in order to determine the most suitable ones for planar conveyance. Finally, a recently introduced approach for the integrated field design and programming of microconveyor platforms for part alignment and orientation is presented.

2.5 References

Akiyama, T. and Shono K (1993a) 'A New Step Motion of Polysilicon Microstructures', Proceedings of IEEE Micro Electro Mechanical Systems Workshop, Fort Lauderdale, Florida: 272–7.

Akiyama, T. and Shono, K. (1993b) 'Controlled Stepwise Motion in Polysilicon Microstructures', *Journal of Microelectromechanical Systems*, 2(3): 106–10.

Ataka, M., Fujita, H. and Mita, M. (2010) '2D Planar Micro Manipulator by Stack-Integrated Micro Actuator/Sensor Array', First Workshop on Hardware and Software Implementation and Control of Distributed MEMS (DMEMS), Besan, TBD, France.

Ataka, M., Mita, M. and Fujita, H. (2007a) 'Multi-Object Conveyance By Peripherally Controlled Micro Actuator/

Sensor Array', *Digest of Technical Papers of the 14th International Conference on Solid-State Sensors, Actuators and Microsystems*, Lyon France: 415–18.

Ataka, M., Mita, M. and Fujita, H. (2007b) *Technical Digest* 'The 20th IEEE International Conference on Micro Electro Mechanical Systems, MEMS2007', Kobe, Japan: 35–8.

Böhringer, K.F. and Donald, B.R. (1996) 'Upper and lower bounds for programmable vector fields with applications to MEMS and vibratory plate parts feeders'. In: International workshop on algorithmic foundations of robotics (WAFR'96), Toulouse, France.

Böhringer, K.F., Donald, B.R. and MacDonald, N.C. (1996) 'What programmable vector fields can (and cannot) do: force field algorithms for MEMS and vibratory plate parts feeders', In: IEEE International Conference on Robotics and Automation (ICRA), Minneapolis, Minnesota: 822–9.

Böhringer, K.F., Suh, J.F., Donald, B.R. and Kovacs, G.T.A. (1997) 'Vector fields for task-level distributed manipulation: experiments with organic micro-actuator arrays', In: IEEE International Conference on Robotics and Automation (ICRA), Albuquerque, New Mexico: 1779–86.

Bronson, J.R., Allen, J.J., Wiens, G.J. (2004) 'Modeling and Characterization of the Scratch Drive Actuator', in: Proceedings of the IMECE: ASME, Anaheim, CA.

Cabrera, S., Harrison, N., Lunking, D., Tang, R., Valentine, T. and Ziegler, C. (2002) 'Latching Shape Memory Alloy Microactuator', ENMA 490 Fall 2002 final report, Department of Materials Science and Engineering University of Maryland. Available from: *http://www.mse.umd.edu/undergrad/490_materials_design/490_fall_2002/490_fall2002_final_project_results/enma490_fall02_final_report_121202.pdf* [accessed 8 April 2011].

Chapuis, Y.A., Fukuta, Y., Mita, Y. and Fujita, H. (2004) 'Autonomous Decentralized Systems Based on Distributed Controlled MEMS Actuators for Micro Conveyance Application', *Seisan-Kenkyu*: 109–16.

Chen, R.S., Kung, C. and Lee, G.B. (2002) 'Analysis of the optimal dimension on the electrothermal microactuator', *Journal of Micromechanical Microengineering*, 12: 291–6.

Dao, D.V., Pham, P.H. and Sugiyama, S. (2008) 'A fully functional micro transportation system with strider-like movement of micro containers', IEEE 21st International Conference on Micro Electro Mechanical Systems MEMS: 50–3.

Delettre, A., Laurent, G.J. and Le Fort-Piat N. (2010) 'A new contactless conveyor system for handling clean and delicate products using induced air flows', IEEE/RSJ International Conference on Intelligent Robots and Systems (IROS): 2351–6.

Denishev, K.H. and Krumova, E.Z. (2005) 'Thermal microactuator', 14th International Scientific and Applied Science Conference ELECTRONICS – ET2005, Sozopol, Bulgaria.

Ebefors, T. (2000) 'Polyimide V-Groove Joints for Three-Dimensional Silicon Transducers', *PhD thesis*, Royal Institute of Technology (KTH), Stockholm, Sweden.

Ebefors, T., Mattsson, J.U., Kälvesten, E. and Stemme, G. (2000) 'A robust micro conveyer realized by arrayed polyimide joint actuators', *Journal of Micromechanical Microengineering*, 10: 337–49.

Egawa, S. and Higuchi, T. (1990) 'Multi-Layered Electrostatic Film Actuator', Proceedings of IEEE Micro Electro Mechanical Systems Workshop, Napa Valley, CA: 166–71.

Egawa, S., Niino, T. and Higuchi, T. (1991) 'Film Actuators: Planar, Electrostatic Surface-Drive Actuators', Proceedings of IEEE Micro Electro Mechanical Systems Workshop, Nara, Japan: 9–14.

Ellis, M. (2004) 'Electrothermal quadmorph microactuator', *US Pat.* 6,679,055, 20 January 2004.

Fearing, R.S. (1998) 'Powering 3 dimensional microrobots: power density limitations', Workshop WS5 on Micromechatronics and Micro Robotics, IEEE International Conference on Robotics and Automation, Leuven, Belgium.

Friend, J., Yeo, L. and Hogg, M. (2008) 'Piezoelectric ultrasonic bidirectional linear actuator for micropositioning fulfilling Feynman's criteria', *Applied Physics Letters*, 92, 014107.

Gill, J.J., Chang, D.T., Momodab, L.A. and Carman, G.P. (2001) 'Manufacturing issues of thin film NiTi microwrapper', *Sensors and Actuators A*, 93: 148–56.

Guckel, H., Klein, J., Christenson, T., Skrobis, K., Laudon, M. and Lovell, E.G. (1992) 'Thermo-Magnetic Metal Flexure Actuators', Solid State Sensors and Actuators Workshop, Hilton Head Island: 73–5.

Judy, J.W. and Muller, R.S. (1996) 'Magnetic microactuation of torsional polysilicon structures', *Sensors and Actuators A: Physical*, 53(1–3): 392–7.

Judy, J.W. and Muller, R.S. (1997) 'Magnetically actuated, addressable microstructures', *Journal of Microelectromechanical Systems*, 6(3): 249–56.

Kanda, T., Makino, A., Suzumori, L., Morita, T. and Kurosawa, M.K. (2004) 'A cylindrical micro ultrasonic motor using a micro-machined bulk piezoelectric transducer', in: IEEE Ultrasonics Symposium, Montreal, Canada: 1298–1301.

Koc, B., Cagatay, S. and Uchino, K. (2002) 'A piezoelectricmotor using two orthogonal bending modes of a hollow cylinder', *IEEE Transactions on Ultrasonics, Ferroelectrics and Frequency Control*, 49: 495–500.

Ku, P.J., Winther, K.T., Stephanou, H.F. and Safaric, R. (2001) 'Distributed control system for an active surface device', Proceedings of the IEEE International Conference on Robotics & Automation, Seoul, Korea: 3417–22.

Lamiraux, F. and Kavraki, L. (2001) 'Positioning of symmetric and non-symmetric parts using radial and constant fields: computation of all equilibrium configurations', *International Journal of Robot Research*, 20(8): 635–59.

Lazarou, P. and Aspragathos, N.A. (2009) 'An integrated mechatronic approach for the systematic design of force fields and programming of microactuator arrays for micropart manipulation', *Journal of Mechatronics*, 19(3): 287–303.

Liua, C., Tsaob, T., Leec, G.B., Leud, J., Yia, Y.W., Taib, Y.C. and Hod, C.M. (1999) 'Out-of-plane magnetic actuators with electroplated permalloy for fluid dynamics control', *Sensors and Actuators A: Physical*, 78(2–3): 190–7.

Luo, J. and Kavraki, L. (2000) 'Part assembly using static and dynamic force fields', In: Proceedings of the IEEE/RSJ International Conference on Intelligent Robots and Systems (IROS): 1468–74.

Mankame, N.D. and Ananthasuresh, G.K. (2001) 'Comprehensive thermal modelling and characterization of an electro-thermal-compliant microactuator', *Journal of Micromechanical Microengineering*, 11: 452–62.

Miyazaki, S., Fu, Y.Q. and Huang, W.M. (2009) *Thin Film Shape Memory Alloys: Fundamentals and Device Applications*. New York: Cambridge University Press.

Moon, I. and Kim, J. (2006) 'Using EWOD (electrowetting-on-dielectric) actuation in a micro conveyor system', *Sensors and Actuators A: Physical*, 130–131: 537–44.

Morita, T., Kurosawa, M.K. and Higuchi, T. (1999) 'Cylindrical microultrasonic motor utilizing bulk lead zirconate titanate (PZT)', *Japanese Journal of Applied Physics*, 38: 3347–50.

Niino, T., Egawa, S., Kimura, H. and Higuchi, T. (1994) 'Electrostatic Artificial muscle: Compact, High-Power Linear Actuators with Multiple-Layer Structures', Proceedings of IEEE Micro Electro Mechanical Systems Workshop, Oiso, Japan: 130–5.

Ren, M.H., Wang, L., Xu, D. and Cai, B.C. (2000) 'Sputter-deposited Ti-Ni-Cu shaped memory alloy thin films', *Materials & Design*, 21(6): 583–6.

Shay, B., Hubbard, T. and Kujath, M. (2008) 'Planar frictional micro-conveyors with two degrees of freedom', *Journal of Micromechanical Microengineering*, 18(6): 065009.

Stemme, G. and Ebefors, T. (2006) Microrobotics, in: *The MEMS Handbook, Vol III, MEMS Applications*, edited by Mohamed Gad-el-Hak. London: Taylor & Francis.

Sudsang, A. and Kavraki, L. (2001) 'A geometric approach to designing a programmable force field with a unique stable equilibrium for parts in the plane', In: Proceedings of the IEEE International Conference on Robotics and Automation (ICRA), Seoul, Korea: 1079–85.

Suh, J.W., Glander, S.F., Storment, C.W., Kovacs, G.T.A. and Darling, R.B. (1997) 'Organic thermal and electrostatic ciliary microactuator array for object manipulation', *Sensors and Actuators A*, 58: 51–60.

Watson, B., Friend, J. and Yeoa, L. (2009) 'Piezoelectric ultrasonic micro/milli-scale actuators', *Sensors and Actuators A: Physical*, 152(2): 219–33.

Wu, C.T. and Hsu, W. (2006) 'Design and fabrication of an electrothermal microactuator for multi-level conveying', *Microsystem Technologies*, 12(4): 293–8.

Xidias, E. and Aspragathos N.A. (2006a) 'A unique stable equilibrium for non-symmetric microparts in the plane',

In: Proceedings of the IWMF Fifth International Workshop on Microfactories, Besancon, France.

Xidias, E. and Aspragathos, N.A. (2006b) 'Force fields with one stable equilibrium for micropart 2D manipulation', In: 4M2006 conference on Multi-Material Micro-Manufacture, Grenoble, France.

Yang, K. and Gu, C.L. (2008) 'Design, optimization and application of novel planar bending ESMAAs', *Journal of Electrical & Electronics Engineering*, 8(1): 519–27.

Zeggari, R., Yahiaoui, R., Malaperta, J. and Manceau, J.F. (2010) 'Design and fabrication of a new two-dimensional pneumatic micro-conveyor', *Sensors and Actuators A: Physical*, 164(1–2): 125–30.

3

Single-axis arm designed with an ultrasonic motor: basic active/passive joint torque control

Fusaomi Nagata, Keisuke Ogiwara and Keigo Watanabe

Abstract: This chapter highlights the active and passive joint torque control methods of a single-axis arm designed with an ultrasonic motor. Recently, many studies on assist robots have been conducted in which the development of a unique system is required to support aged persons, physically handicapped persons and/or carers. One of the representative systems is called the assist suit and is partially developed. The assist suit is a mechatronics device which can assist physical human actions. However, the current assist suit has a few problems with respect to cost, size, weight, longevity and so on. In this chapter, a fundamental study concerning a compact assist device is conducted. Where the assist device supports is one spot on the body such as a knee, an elbow or a shoulder. First of all, a simple single-axis arm is designed by using an ultrasonic motor which can generate high torque from a low velocity range. Then, a servo system with holding torque, a torque control system and a passive torque control system are proposed and their characteristics are evaluated. Here, the passive torque control includes a stiffness control, a compliance control and an impedance control.

Key words: ultrasonic motor; single-axis arm; servo system; holding torque; torque control; compliance control; assist device.

3.1 Introduction

Recently, many studies on assist robots have been conducted in which the development of a unique system is required to support aged persons, physically handicapped persons and/or carers (Toyama and Yonetake, 2007). One of the representative systems is called the assist suit and is partially developed. The assist suit is a mechatronics device which can assist physical human actions. However, the current assist suit has a few problems with respect to cost, size, weight, longevity and so on.

In this chapter, a fundamental study concerning a compact assist device is conducted (Ogiwara and Nagata, 2010). Where the assist device supports is one spot on the body such as a knee, an elbow or a shoulder. First of all, a simple single-axis arm is designed by using an ultrasonic motor which can generate high torque from a low velocity range. Figure 3.1 shows the experimental set-up. Then, a servo system by using holding torque, a joint torque control system and a passive torque control system are proposed and their characteristics are evaluated. The passive torque control includes a stiffness control, a compliance control and an impedance control. The proposed passive torque control is easily realised by using the servo system as an inner loop, comparing with the position-based impedance

Figure 3.1 **Single-axis arm designed with an ultrasonic motor**

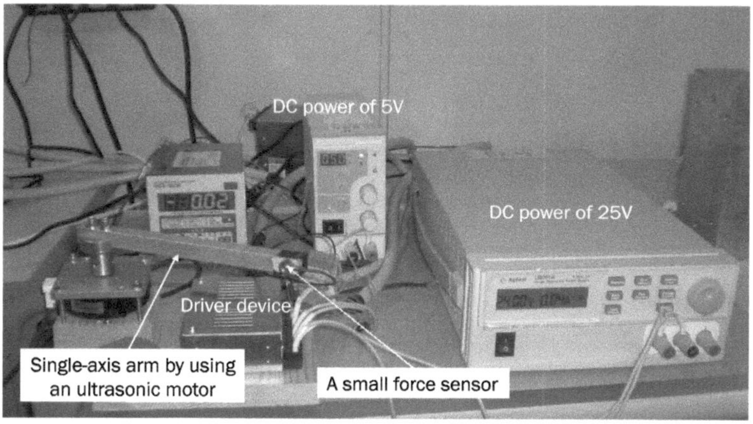

control (Heinrichs and Sepehri, 1997, 1999). Finally, a promising application as an assist device is considered. It is assumed that the single-axis arm is fitted to a damaged joint of a human. The operator can adjust the desired joint torque, e.g. more stiff or more compliant, while assessing the behavior and the effectiveness of the assist device.

3.2 Single-axis arm designed with an ultrasonic motor

The ultrasonic motor has two features. One is that it can generate high torque from the low velocity range. The other is that it has a large holding torque when no voltage is given. That is the reason why no brake system is needed and consequently the weight reduction can be realised. Also, the responsiveness of the ultrasonic motor is generally superior to the one of conventional electromagnetic motors, so that energy consumption can be suppressed. Considering the above points, the simple single-axis arm is designed based on an ultrasonic motor (Ogiwara et al., 2011). Figure 3.2 shows the hardware block diagram. The ultrasonic motor used in experiments is the model of USR60-E3T provided by Shinsei Corporation. A DA board (CONTEC DA12-8) is used to control the motor velocity. A digital IO board (CONTEC PIO-48D) is also used to switch the direction of motor rotation, i.e., clockwise or counterclockwise. Further, a counter board (CONTEC CNT32-8M) is incorporated to sense the rotation angle. These cards are connected to the ultrasonic motor via a driver device (Shinsei Corp. D6060 24V). A small force sensor is fixed at the tip of the single-axis arm to estimate the joint torque and another small force sensor is used as a simple input device for an operator.

Figure 3.2 Hardware block diagram of the experimental system

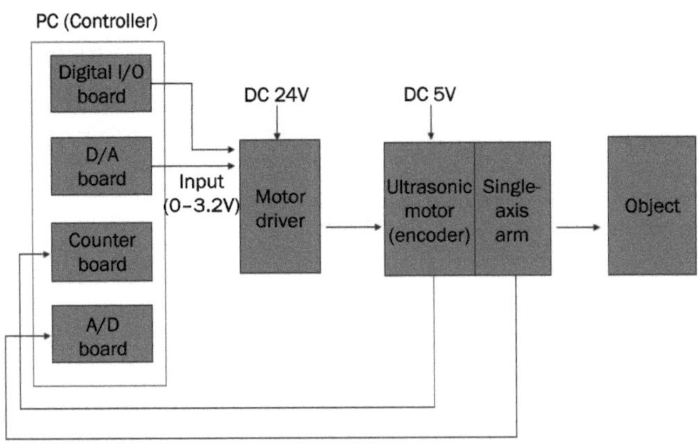

3.3 Control system

Basic functions were first developed on Microsoft Visual C++ to give the commands such as the direction of rotation and the rotational velocity and to obtain the rotational angle from the encoder. Then, it was measured on how the relation between the voltage and its given time influenced the dynamic response of the motor. Figure 3.3 shows the characteristics, in which it is confirmed that the velocity tends to be constant under 0.6V. The dynamic characteristics are used to design a position feedforward controller in the next subsection.

Figure 3.3 Relation between joint driving torque and rotational speed in steady state

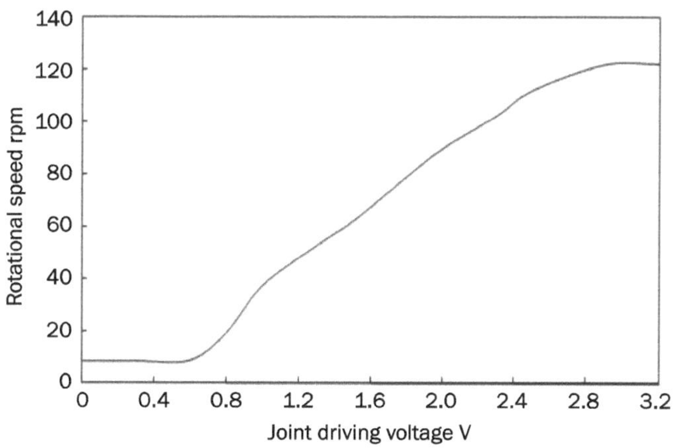

3.3.1 Basic servo system

The servo system of an ultrasonic motor is easily constructed due to the high holding torque and the responsiveness. Here, a simple proportional control has only to be applied as:

$$\tau(k) = K_p\left[\theta_d - \theta(k)\right] \tag{3.1}$$

where $\tau(k)$ is the joint driving voltage at the discrete time k, K_p is the p-gain, $\theta(k)$ and θ_d are the joint angle and desired one, respectively. In order to conduct high-accuracy positioning, when $\theta(k) = \theta_d$ is detected in the sampling loop, the excitation power to the motor has only to be off at the same time. Figure 3.4 shows examples of step response with several K_p, in which 500 pulses mean π rad. Note that a linear characteristic suddenly appears from a point, for example, in case of $K_p = 0.006$. To examine the phenomenon in a bit more detail, the relation between the time and the torque obtained by Equation 3.1 is measured as shown in Figure 3.5. It is observed from Figures 3.3, 3.4 and 3.5 that the joint velocity tends to show a constant value about 8.5rpm under the point of 0.6V. So, in order to cope with the motor characteristics, Equation 3.1 is improved as

$$\tau(k) = K_p\left[\theta_d - \theta(k)\right] + 0.6 \tag{3.2}$$

where 0.6 means the feedforward quantity. Figure 3.6 shows the step response when Equation 3.2 is employed. As can be seen, a desirable response without an overshoot and a large delay is observed only by using a p-action. This is the

Figure 3.4 **Step responses obtained by using Equation 3.1, in which 500 pulses mean π rad**

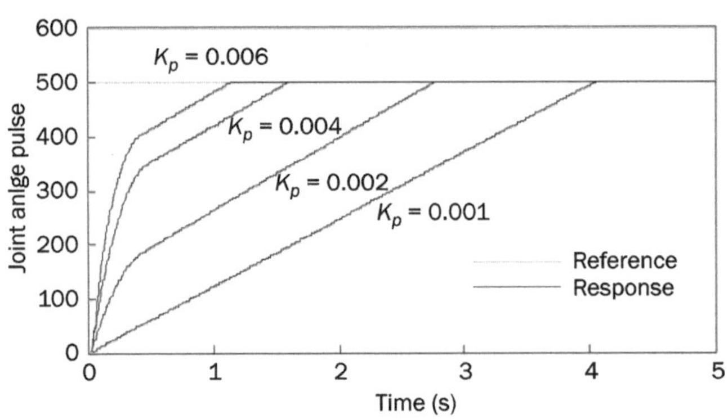

Figure 3.5 Joint driving voltage calculated by using Equation 3.1

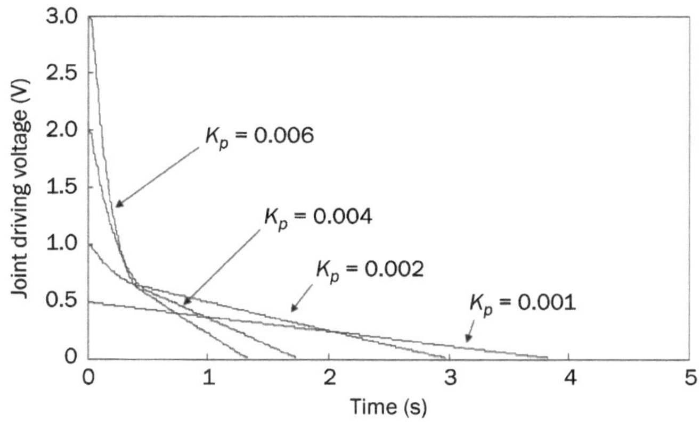

Figure 3.6 Step responses obtained by using Equation 3.2

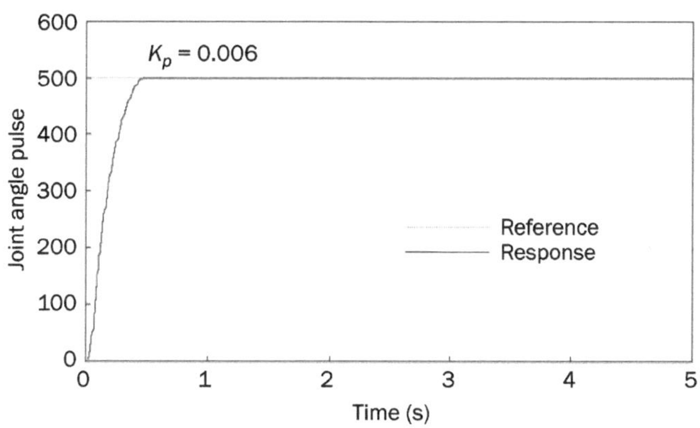

attractive characteristics of the ultrasonic motor used in the experiment.

3.3.2 Joint torque control

In the joint torque control mode, the torque acting at the joint is actively controlled by a PI controller designed as

$$\tau(k) = K_{fp}\left[\tau_d - \tau_s(k)\right] + K_{fi}\sum_{n=1}^{k}\left[\tau_d - \tau_s(n)\right] + 0.6 \qquad (3.3)$$

where K_{fp} and K_{fi} are the p-gain and i-gain, respectively. τ_d is the desired joint torque, $\tau_s(k)$ is the estimated joint torque which is calculated from the force value sensed by a small force sensor attached to the arm tip. Figure 3.7 shows a torque control result, in which the response desirably follows the reference 2Nm by setting K_{fp} = 3 and K_{fi} = 0.001, respectively.

Figure 3.7 Joint torque control result by using Equation 3.3

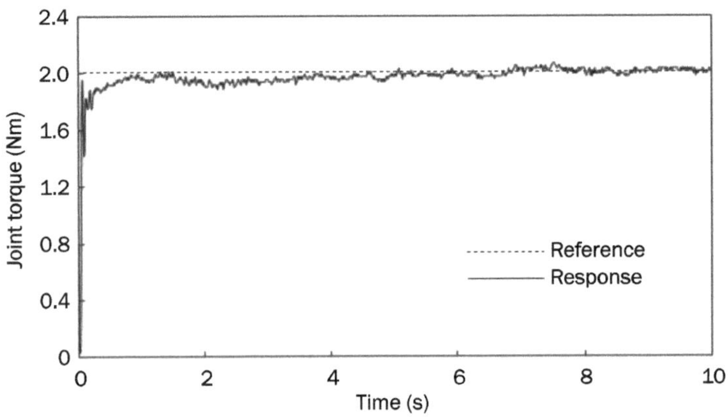

3.3.3 Passive joint torque control

In the passive joint torque control mode, an external force given to the arm can be absorbed smoothly. Here, a stiffness control, a compliance control and an impedance control are considered by using the inner servo system given by Equation 3.2.

Stiffness control

A stiffness control law is designed as

$$\tau_s(k) = K_d\left[\theta(k) - \theta_d\right] \tag{3.4}$$

where K_d is the desired stiffness [Nm/rad]. Note that in the stiffness control mode, the initial position is set to the desired position θ_d in Equation 3.4, and also $\theta(k)$ obtained from Equation 3.4 is given to θ_d in Equation 3.2, so that the stiffness control law is represented by

$$\tau(k) = K_p\left(\frac{\tau_s(k)}{K_d} - \theta(k)\right) + 0.6 \tag{3.5}$$

where 0 is given to θ_d in Equation 3.4. Figure 3.8 shows stiffness control results with five values of K_d.

Figure 3.8 **Stiffness control result by using Equation 3.5**

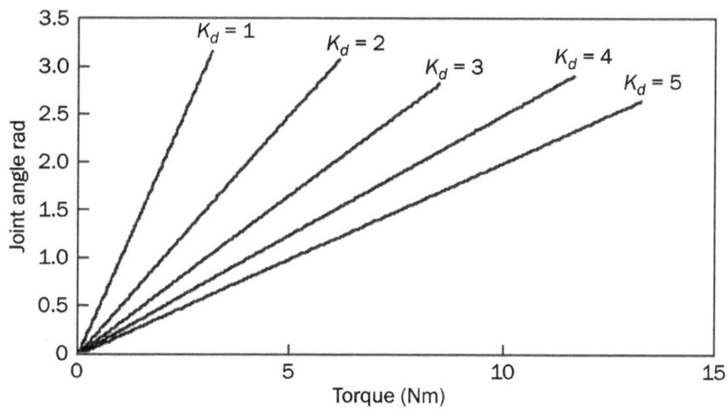

Compliance control

Next, a compliance control law is designed as

$$\tau_s(k) = B_d\left[\dot{\theta}(k) - \dot{\theta}_d\right] + K_d\left[\theta(k) - \theta_d\right] \tag{3.6}$$

where B_d is the desired viscosity [Nm·s/rad]. In the compliance control mode, the transient behaviour to an equilibrium position can be controlled through a one-order lag system. Of course, the equilibrium position depends on K_d. If it is assumed that both $\dot{\theta}_d$ and θ_d are 0, and $\tau_s(k) = \tau_{step}$ (constant), then the solution $\theta(k)$ of Equation 3.6 is given by

$$\theta(k) = \frac{\tau_{step}}{K_d}\left(1 - e^{-\frac{K_d}{B_d}\Delta tk}\right) \tag{3.7}$$

where Δt is the sampling width. The compliance control can be easily realised by giving $\theta(k)$ obtained from Equation 3.7 into θ_d in Equation 3.2, which is given by

$$\tau(k) = K_p\left[\frac{\tau_{step}}{K_d}\left(1 - e^{-\frac{K_d}{B_d}\Delta tk}\right) - \theta(k)\right] + 0.6 \tag{3.8}$$

Figure 3.9 shows examples of compliance control results, in which the transient behaviours are changed with five B_d. In the experiment, K_d and τ_{step} are set to 1 and 2, respectively.

Figure 3.9 Compliance control result by using Equation 3.8

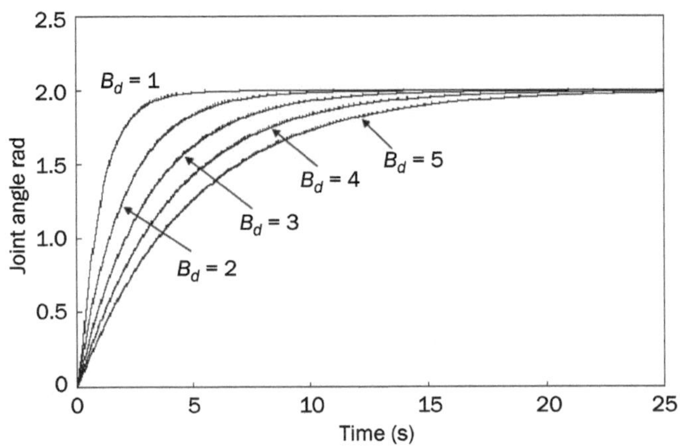

Impedance control

An impedance control method is further designed by

$$\tau_s(k) = M_d\left[\ddot{\theta}(k) - \ddot{\theta}_d\right] + B_d\left[\dot{\theta}(k) - \dot{\theta}_d\right] + K_d\left[\theta(k) - \theta_d\right] \quad (3.9)$$

where M_d is the desired inertia [kg·m^2]. $\ddot{\theta}(k)$ and $\dot{\theta}_d$ are the joint acceleration and desired one, respectively. In the impedance control mode, the transient behaviour to an equilibrium position can be controlled as a second-order lag system with under-damped condition. The equilibrium position depends on K_d, too. If it is assumed that $\ddot{\theta}_d$, $\dot{\theta}_d$ and θ_d are set to 0, and $\tau_s(k) = \tau_{step}$ (constant), then the solution $\theta(k)$ of Equation 3.9 is derived as

$$\theta(k) = \frac{\tau_{step}}{K_d}\left[1 - \frac{e^{-\zeta\omega_n t}}{\sqrt{1-\zeta^2}}\sin\left(\sqrt{1-\zeta^2}\,\omega_n t + \tan^{-1}\frac{\sqrt{1-\zeta^2}}{\zeta}\right)\right] \quad (3.10)$$

where the damping coefficient ζ and natural frequency ω_n are given by

$$\zeta = \frac{B_d}{2\sqrt{M_d K_d}} \quad (3.11)$$

$$\omega_n = \sqrt{\frac{K_d}{M_d}} \quad (3.12)$$

Note that the condition of $0<\zeta<1$ must be given to make the system being under damped. By substituting Equation 3.10 into θ_d in Equation 3.2, the impedance control can be easily realised by using the servo system as

$$\tau(k) = K_p\left\{\frac{\tau_{step}}{K_d}\left[1 - \frac{e^{-\zeta\omega_n t}}{\sqrt{1-\zeta^2}}\sin\left(\sqrt{1-\zeta^2}\,\omega_n t + \tan^{-1}\frac{\sqrt{1-\zeta^2}}{\zeta}\right)\right] - \theta(k)\right\} + 0.6 \quad (3.13)$$

Figure 3.10 Impedance control result by using Equation 3.13

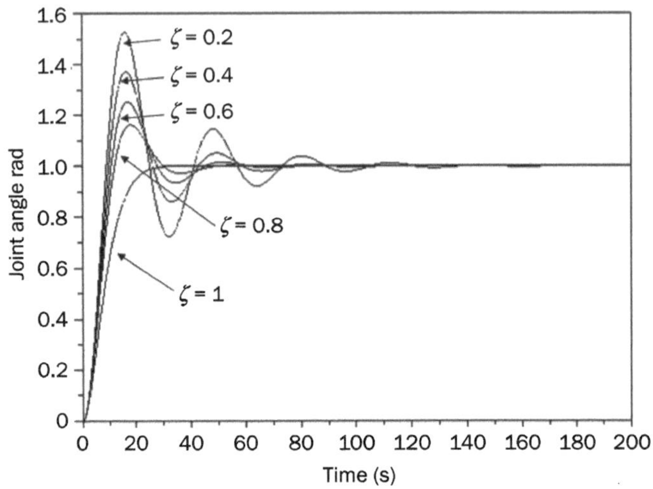

Figure 3.10 shows impedance control results by using Equation 3.13. It is confirmed that the responses desirably vary according to five values of damping coefficients ζ.

3.4 Example of application

In this section, a promising application called an assist device is introduced. It is assumed that the single-axis arm is fitted to a damaged joint of a human. A small and lightweight force sensor is used so that the desired torque can be directly given by the fingers of the operator. The operator can adjust the desired torque, e.g. more stiff or more compliant, while assessing the behaviour and the effectiveness of the assist device, i.e., the single-axis arm.

The block diagram of the controller is shown in Figure 3.11, in which the desired torque $\tau_d(k)$ is manually given by an operator. The control law given by Equations 3.14, 3.15

Figure 3.11 **Block diagram of an application called the assist device for assisting a damaged or weakened joint**

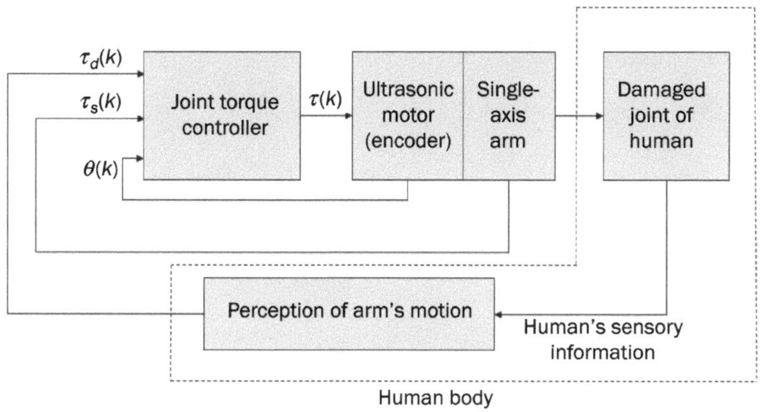

and 3.16 is improved based on Equation 3.3, in which the desired joint torque is time-varying.

$$\tau(k) = K_{fp}\left[\tau_d(k) - \tau_s(k)\right] \tag{3.14}$$

$$+K_{fi}\sum_{n=1}^{k}\left[\tau_d(k) - \tau_s(n)\right] + 0.6$$

$$\tau_d(k) = \alpha f(k) \tag{3.15}$$

$$\tau_s(k) = LF(k) \tag{3.16}$$

where $f(k)$ is the small force generated by an operator's fingers, α is the gain which transforms the force to the time-varying desired joint torque $\tau_d(k)$. $F(k)$ is the force acting between the arm tip and the object, L is the length of the single-axis arm. In the experiment, $\tau_s(k)$ is regarded as the joint torque.

Figure 3.12 shows the experimental scene assumed to be the assist device. The control result of $\tau_s(k)$ is shown in Figure 3.13, in which the response $\tau_s(k)$ is desirably amplified according to $f(k)$. In this case, 1 and 0.15 are given to α and L, respectively.

Figure 3.12 Experimental scene assumed to be the assist device

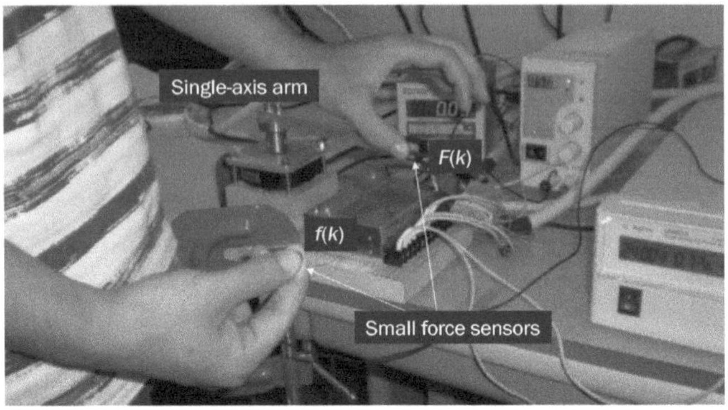

Figure 3.13 Joint torque manually controlled by an operator, in which the force is given by the operator's fingers

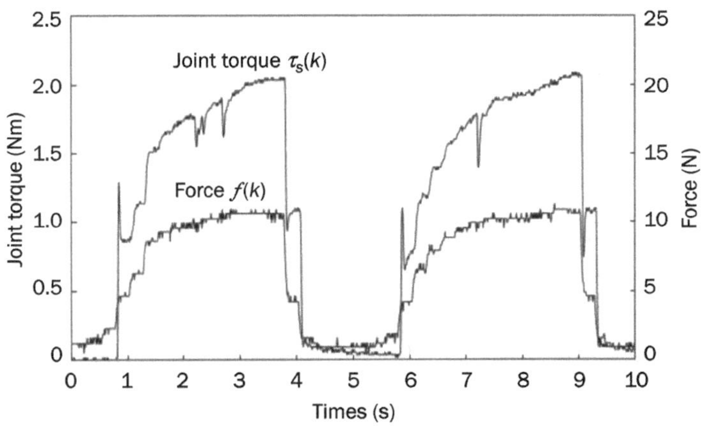

3.5 Conclusions

In this chapter, a fundamental study concerning a compact assist device is conducted. Where the assist device supports is one spot on the body such as a knee, an elbow or a shoulder. First of all, a simple single-axis arm has been designed by using an ultrasonic motor which can generate high torque from a low velocity range. Then, a servo system by using holding torque, a joint torque control system and a passive torque control system have been applied and their characteristics are evaluated. The passive torque control includes a stiffness control, a compliance control and an impedance control. Finally, a promising application as an assist device has been considered. It is assumed that the single-axis arm is fitted to a damaged joint of a human. The operator can adjust the desired torque, e.g., more stiff or more compliant, while assessing the behaviour and the effectiveness of the assist device.

3.6 References

Heinrichs, B. and Sepehri, N. (1997) 'Position-based impedance control of an industrial hydraulic manipulator,' *IEEE Control Systems Magazine*, 17(1): 46–52.

Heinrichs, B. and Sepehri, N. (1999) 'A limitation of position based impedance control in static force regulation: theory and experiments,' *Proceedings of 1999 IEEE International Conference on Robotics and Automation*, 3: 2165–70.

Ogiwara, K. and Nagata, F. (2010) 'Basic position/force control of a single-axis arm designed with an ultrasonic motor,' *Proceedings of the 12th IEEE Hiroshima Section Hiroshima Student Symposium*, (F)–11, 2 pages, (in Japanese).

Ogiwara, K., Nagata, F. and Watanabe, K. (2011) 'Basic position/force control of single-axis arm designed with an ultrasonic motor,' *Proceedings of the 16th International Symposium on Artificial Life and Robotics*: 383–6.

Toyama, S. and Yonetake, J. (2007) 'Development of the ultrasonic motor-powered assisted suit system,' *Proceedings of 2007 IEEE/ICME International Conference on Complex Medical Engineering*: 1361–6.

4

Signal processing for tool condition monitoring: from wavelet analysis to sparse decomposition

Zhu Kunpeng, Wong Yoke San and Hong Geok Soon

Abstract: Signal processing plays an important role in manufacturing automation and industrial control. In CNC machining, by utilising the monitored information from the platform to direct the further actions, the signal processing bridges the gap of human instruction and full automation. One of the main problems related to CNC machining productivity and safety is the tool conditions. Effective signal processing techniques are prerequisite for tool condition monitoring (TCM). Fourier transform is the basis of modern signal processing and has been the main signal processing method. It associates the signal from the time domain to the frequency domain and provides useful information. Wavelet analysis is inherited from Fourier analysis and is most suitable for analysing non-stationary signal processing in machining. This chapter serves as an introduction to the wavelet analysis and the newly developed sparse decomposition approaches, with regard to their applications to TCM.

Key words: tool condition monitoring; wavelet analysis; sparse decomposition.

4.1 Overview of tool condition monitoring and signal processing issues

4.1.1 Overview of tool condition monitoring

CNC machining is fundamental to modern manufacturing. During machining, the contact between the cutting tool, workpiece and the chips imposes pressure on the tool and causes the shape of the tool to change, either gradually as tool wear or abruptly as tool fracture or breakage. Tool abnormal conditions can range from slightly exceeding the tolerances specified for the product to catastrophic failure of a tool which destroys an extremely expensive workpiece. The aim of tool condition monitoring is to apply appropriate sensor signal processing and artificial intelligence (AI) techniques to identify and predict the cutting tool state, so as to reduce loss brought about by machining failures. An effective TCM system can improve productivity, safety and ensure product quality, and hence has a major influence on machining efficiency (Byrne et al., 1995). Tool condition monitoring has been extensively studied by many researchers since the late 1980s. There are vast literatures, and many of the reported research works are reviewed in Byrne et al. (1995), Liang et al. (2004) and Zhu et al. (2009b).

Basically, the TCM system can be achieved in a three-step procedure as a pattern recognition problem (Figure 4.1). Under this framework, the objective of TCM is to search the most probable state C_i given the extracted measurable signal feature $y(t)$ at time t. The sensor signal $x(t)$ is firstly pre-processed to remove the noise and prepare the data for feature extraction. Then the information relevant to pattern classification is extracted from $x'(t)$ to a feature vector $y(t)$. The task of feature extraction is to enhance the characteristics of the various tool wear classes and suppress or filter off the normal background. The final stage is state classification. Feature vector $y(t)$ is assigned to one of the K tool wear state, C_1, C_2, . . . C_K, by the classifier based on a certain type of classification criteria. Practically indirect measurement methods such as force, acoustic emission, and vibration are used to capture machining signals. After the signal's pre-processing and feature extraction, approaches such as autoregressive models (Liang and Dornfeld, 1989), neural networks (Tansel et al., 1998; Wang et al., 2007), and hidden Markov models (HMMs) (Zhu et al., 2009a) are then applied to estimate the tool wear states.

Figure 4.1 TCM as a pattern recognition system

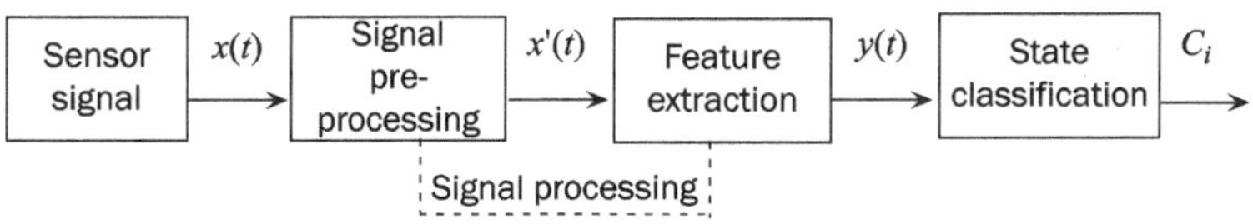

4.1.2 Signal processing issues in TCM

In spite of the sophisticated classifiers, the signal processing part is most important for a successful TCM system. For successful implementation of a TCM system, the sensor signals should be sensitive to the tool conditions in a harsh machining environment. Sensor signals are typically noisy and have to be pre-processed to yield useful features that are highly correlated to tool conditions. This process is called signal pre-processing and feature extraction. Typical procedures in signal pre-processing are de-noising and normalisation, which steps are applied to clear the signal and eliminate the effect of different dynamic ranges because of different cutting conditions.

The feature extractions in TCM have been extensively studied. Earlier works focused on time series methods and frequency domain analyses, such as autoregressive (AR) time series model developed by Liang and Dornfeld (1989), fast Fourier transform (FFT) employed by Elbestawi et al. (1991) and Tarng et al. (1993). The AR and FFT approaches inherently assume that the signals are stationary. However, due to the nature of manufacturing processes, the metal cutting processes involve a large number of transient, the signals are usually non-stationary. Thus, the approaches that deal with non-stationary signals are more appropriate for process monitoring. Wavelet transform is a convenient tool to meet the needs. It has been applied to analyse machining signals during tool wear by Tansel et al. (1998), Lee and Tarng (1999), Zhu et al. (2008a, 2009a). A comprehensive review of apply wavelet approaches to TCM may be found in Zhu et al. (2009b).

These signal processing methods, with their advanced mathematical theory, are promising and could extract useful information for TCM. Unfortunately, the academic signal processing work to date is predominantly oriented towards other applications with traditionally stronger ties to the signal processing community, or to theoretical issues which are far beyond the scope of mechanical engineering. The purpose of this chapter is to bring the academic researchers and the industrial users together to identify the promising signal processing technologies that meet the demands of TCM. However, an exhaustive presentation of these vast topics would be beyond the scope and possibilities of a single chapter. The signal processing approaches discussed in this chapter are expressed with linear transforms, from Fourier transform to sparse decomposition, with a focus on wavelet transform. The outline of the chapter is as follows: first, some basic concepts of signal processing are reviewed, including signal space, the Fourier transform and sampling. Then, classical spectral analysis methods based on the Fourier transform and their limitations will be discussed. In the following section, the theory of wavelet analysis is introduced and discussed with more details. The theory of sparse signal representation is then introduced. In the last part of the chapter, some of the approaches' applications to TCM are presented.

4.2 Signal space, linear system and Fourier transform

For a fundamental understanding of signal theory, a basic knowledge of functional analysis is needed, such as the definition of Hilbert space, metric space, frame and operator theory. These concepts are far beyond the scope of this chapter, and we only introduce the theory related to Fourier and wavelet analysis. The theories presented herein are not mathematically rigorous but are aimed at easy understanding. Interested readers may refer to Franks (1969) on theory of signal analysis and Moon and Stirling (2000) for a modern approach from a vector space view.

4.2.1 Signal spaces and inner product

In a broad sense, a signal $f(t)$ is defined as a time-varying function of time variable t,

$$\text{Signal} = \text{function } f(t), t \in R \tag{4.1}$$

A signal space is simply a collection of signals (functions) that satisfies a certain mathematical structure. The signal spaces with finite energy and finite power structures are particularly interesting in signal processing.

Finite energy space $L_2(R)$ is the vector space of finite energy functions (or say, square integrable) such that,

$$\int_{-\infty}^{\infty} |f(t)|^2 \, dt < +\infty \tag{4.2}$$

The signal in this space is called energy signal. For a signal $f(t)$, if the energy is infinite, but the mean power is finite in the time period (-T/2, T/2) in finite,

$$\frac{1}{T}\int_{-T/2}^{T/2} |f(t)|^2 \, dt < +\infty \tag{4.3}$$

it is called power signal. Most signals encountered in technical applications belong to these two classes.

If $f(t)$ is an energy signal, then we can consider $f(t)$ as belonging to the Hilbert space, $L_2(R)$. The reader unfamiliar with Hilbert space may think of it as a generalisation of the three-dimensional Euclidean space, in which the concepts of distance and angle are extended as norm and inner product respectively. For any signal in the Hilbert space, it has the basic operation properties that can be added, subtracted and multiplied by constants.

The inner product of two signals $x(t)$ and $y(t)$ is defined as

$$\text{inner product} = \langle x(t), y(t) \rangle = \int_{-\infty}^{\infty} x(t) y^*(t) \, dt \tag{4.4}$$

where $y^*(t)$ is the complex conjugate of $y(t)$, and if $y(t)$ is real then $y^*(t) = y(t)$. The inner product is a generalisation of dot product in the signal (vector) space.

The $x(t)$ and $y(t)$ are said to be orthogonal if their inner product $\langle x(t), y(t)\rangle = 0$, and if each vector is also a unit vector, the vector set is called orthonormal. The collection of these vectors can span completely any vector in the space. They are called the orthogonal (orthonormal) basis. The inner product can be roughly understood as correlation of $x(t)$ and $y(t)$: the higher their correlation, the larger the absolute value. If they are uncorrelated, then $\langle x(t), y(t)\rangle = 0$, $x(t)$ and $y(t)$ are orthogonal.

In the Hilbert space, we can always find a set of functions $\{f_i(t)\}_{i=1}^{\infty}$ in $L_2(R)$ that are orthonormal, i.e., $\langle f_i(t), f_j(t)\rangle = 0$, *when* $i \neq j$, such that any $s(t) \in L_2(R)$, we have,

$$s(t) = \sum_{i=1}^{\infty} s_i f_i(t) \tag{4.5}$$

The set $\{f_i(t)\}_{i=1}^{\infty}$ is called a complete basis for $L_2(R)$. This leads to the signal's linear decompositions with basis for the later parts of this chapter.

4.2.2 Fourier transform

For any signal $f(t)$, the Fourier transform $\hat{f}(\omega)$ is obtained by inner product of $f(t)$ with a sinusoidal wave $e^{j\omega t}$,

$$\hat{f}(\omega) = \left\langle f(t), e^{j\omega t} \right\rangle = \int_{-\infty}^{\infty} f(t) e^{-j\omega t}\, dt \tag{4.6}$$

The FT transforms the signal $f(t)$ from the time domain into the frequency domain, and results in the amplitudes of the frequency ω.

There is another equation that reverses the transform and takes the signal from the frequency domain back to the time domain,

$$f(t)=\frac{1}{2\pi}\int_{-\infty}^{\infty}\hat{f}(\omega)e^{j\omega t}\,d\omega \tag{4.7}$$

Equations 4.6 and 4.7 can be used on any signal, periodic or non-periodic. The FT lays down the basis of modern signal processing. It was extensively applied to TCM in earlier works, for example Elbestawi et al. (1991), Tarng et al. (1993).

4.2.3 Linear system, sampling theorem and convolution

The Fourier transform is powerful and dominates signal processing because the sinusoidal waves $e^{-j\omega t}$ are the eigenvectors of time invariant linear system. A system or transform maps an input signal $x(t)$ into an output signal $y(t)$,

$$y(t)=T\left[x(t)\right] \tag{4.8}$$

where T denotes the system or transform, a function from input signals to output signals. Systems come in a wide variety of types.

One important class is known as linear systems. A linear system obeys rules of linearity, which are homogeneity and additivity. Specifically, a linear system T that maps a vector space X into a vector space Y is said to be linear if for every scalar α, β we have,

$$T(\alpha x+\beta y)=\alpha T(x)+\beta T(y) \tag{4.9}$$

Please note that the linear system here refers to the signal processing system, not to a claim that our machining system is linear. Linear signal processing brings a lot of good properties and is favourable with signal analysis. Fourier analysis, wavelet analysis and sparse decomposition are all linear. The non-linear time frequency approaches such as Wigner-Wille distribution (Cohen, 1989) are not discussed in this chapter.

In practice, the continuous signals are sampled at discrete time to process. It can be completely represented by a set of equally spaced samples, if the samples occur at more than twice the frequency of the highest frequency component of the signal. This sampling frequency is called the Nyquist frequency (Shannon, 1949). If we sample the signal $x(t)$ with sampling frequency f_s (sampling interval $T_s = 1/f_s$), then we generate the sequence $\{\ldots, x(-nT_s), \ldots, x(-T_s), x(0), x(T_s), \ldots, x(nT_s), \ldots\}$,

$$y(t) = \sum_{n=-\infty}^{\infty} x(nT_s) \cdot \delta(t - nT_s) = x(t) \cdot \sum_{k=-\infty}^{\infty} \delta(t - nT_s) \qquad (4.10)$$

where $\delta(t - nT_s) = 1$ at $t - nT_s$ and equals 0 elsewhere. The maximum seeable frequency in the $X(\omega)$ by DFT is $f_s/2$ according to this theory.

The Nyquist frequency has dominated the sampling principles in signal processing, but it has been discovered lately (Candès et al., 2006) that this constraint is not necessary when the signal is sparse in some domain (either time, frequency, or another). This finding leads to the newly developed theory called sparse decomposition (or compressive sensing/sampling in information theory), which is to be discussed in 'Sparse coding' below.

Convolution is another important concept in signal processing. For a linear system, let $h(t)$ be the impulse response (the output of the system to the input of delta function) of the system, then for any input $x(t)$, the output $y(t)$ is obtained by convolution,

$$y(t) = Lx(t) = h(t) * x(t) = \int_{-\infty}^{\infty} h(\tau) x(t - \tau) \, d\tau \qquad (4.11)$$

then the system is called linear.

The convolution has a clear physical meaning that: at any time, the system's output $y(t)$ is obtained by the input $x(t)$ and the convolution of the impulse response of the system $h(t)$. And if $y(t - t_0) = Lx(t - t_0)$, any input delay will result in an output delay, the system is called linear time invariant system (LTI). Convolution has a special importance in the understanding of wavelet. As a matter of fact, Mallat and Hwang (1992) defined the wavelet transform by convolution.

4.3 Wavelet analysis

4.3.1 Limitation of Fourier methods: time-frequency resolutions of STFT

In spite of its popularity, Fourier transform has certain serious theoretical drawbacks in processing signals, which limits it applications to TCM. This is because the support of $e^{j\omega t}$ covers the entire real dimension (Equation 4.6) so that $\hat{f}(\omega)$ depends on the value of $f(t)$ for all times $t \in R$. This global inclusivity of information makes it difficult to analyse any local property of $f(t)$ from $\hat{f}(\omega)$ with WT. As can be seen from Figure 4.2, it has infinite time resolution (upper Figure 4.2a), and has infinite frequency resolution (Figure 4.2b), but it does not have the information regarding the time-frequency relationship. To overcome the limitation of the lack of localisation, Gabor (1946) introduced a sliding window function $g(t)$ to the Fourier transform and obtains a localised time-frequency atom ϕ:

$$\phi_{u,\xi}(t) = e^{j\xi t} g(t-u) \tag{4.12}$$

The resultant transform is named short-time Fourier transform (STFT).

$$(STFT)(f(t)) = \langle f(t), \phi_{u,\xi} \rangle = \int_{-\infty}^{\infty} f(t)g(t-u)e^{-j\xi t}dt \qquad (4.13)$$

The corresponding energy density $|STFT(f(t))|^2$ is called a spectrogram, which was widely used for time-frequency analysis before wavelet dominated the area in the 1990s, and applied to TCM by Dowling (1993). Figure 4.2c is the spectrogram of the force from Figure 4.2a. The power spectrum is localised in the time axis and the variation of spectrum energy can be detected in the spectrogram.

In the STFT analysis, however, once the window is chosen, the frequency resolution σ_ω^2 and time resolution σ_t^2 are constant for both time and frequency, and according to the Heisenberg Uncertainty Principle (Mallat, 2008),

$$\sigma_t^2 \sigma_\omega^2 \geq \frac{1}{4} \qquad (4.14)$$

equality holds if $\phi(t)$ is a Gaussian.

It states that Fourier methods cannot reach high resolution both in frequency and time. On the other hand, the above-mentioned approaches assume that the sensor signals are stationary. However, due to the nature of manufacturing processes, the signals are usually non-stationary (Wu and

Figure 4.2 Cutting forces, Fourier transform and spectrogram

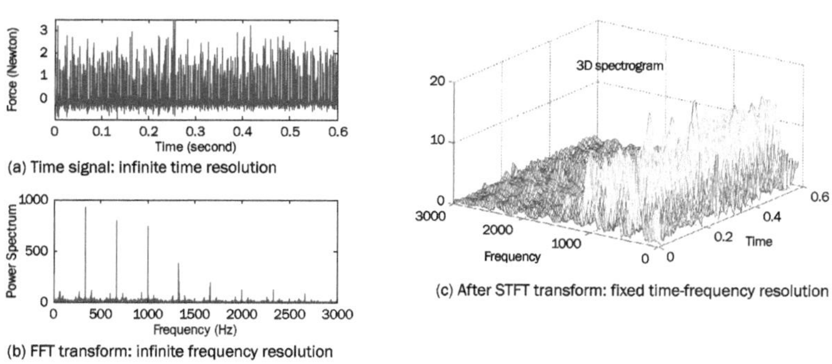

Du, 1996). Constant time and frequency resolutions of STFT are not suitable for the analysis of non-stationary signal. For example, Mori et al. (1999) took the FFT of the thrust force signal and found that the spectra cannot capture the time-localised aspect of the saw-tooth signal, but instead it spread the information across the transformed signal. The signal processing approaches that deal with non-stationary signals are more appropriate for process monitoring. Gong et al. (1997) have shown that the wavelet analysis is more sensitive and reliable than the Fourier analysis for recognising the tool wear states in turning. Yoon et al. (2005) also verified the reliability of the wavelet transform method compared with the spectra method of FFT. Wavelet analysis overcomes the drawbacks of Fourier methods and permits adaptive time-frequency representation.

4.3.2 Continuous wavelet analysis (CWT) and its time-frequency properties

Continuous wavelet analysis (CWT)

To meet the needs for adaptive time-frequency analysis in applied mathematics, physics and engineering, the wavelet theory was developed in the late 1980s by Morlet, Meyer, Mallat and Daubechies (see Mallat, 1989; Daubechies, 1990). In wavelet transform (WT), a signal of interest is decomposed into a set of basic localised waveforms, called wavelets. The signal is analysed by examining the coefficients of the wavelets. Wavelet transform is a convenient tool for processing non-stationary signals. It has been widely used to analyse machining signals for tool condition monitoring.

Let $\psi_{a,b}(t)$, a, $b \in R$, $a \neq 0$, be a family of functions defined as translations and re-scales of a single function $\psi_{a,b}(t) \in L^2(R)$ (Mallat, 1989),

$$\psi_{a,b}(t) = \frac{1}{\sqrt{|a|}} \psi\left(\frac{t-b}{a}\right) \tag{4.15}$$

with the following basic properties,

$$\int_R \psi_{a,b}(t)dt = 0 \tag{4.16}$$

$$\int_R \psi_{a,b}^2(t)dt = 1 \tag{4.17}$$

These properties indicate that the wavelet is a small wave: oscillating around zero (zero mean) and diminishing quickly (finite energy).

But not all $\psi_{a,b}(t) \in L^2(R)$ with the properties in Equations 4.16 and 4.17 can be regarded as wavelet. The wavelet $\psi_{a,b}(t)$ has to meet the following *admissibility condition* for the transformation to be invertible (to reconstruct the signal after WT transform),

$$C_\psi = \int_{-\infty}^{\infty} \frac{\left|\hat{\Psi}(\omega)\right|^2}{\omega} d\omega < \infty \tag{4.18}$$

Figure 4,3 shows samples of Haar, Morlet and Daubechies wavelets.

The continuous wavelet transform (CWT) of signal $f(t) \in L^2\ (R)$ is defined as the inner product of $f(t)$ and $\psi_{a,b}(t)$,

Figure 4.3 **Haar wavelet, Morlet wavelet and Daubechies wavelet**

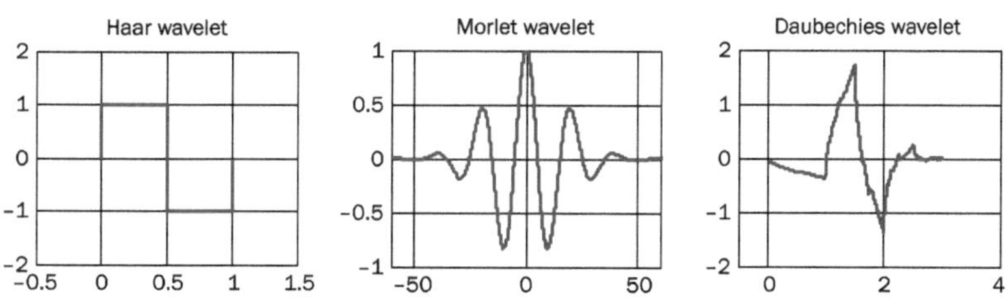

$$W_\psi f(a,b) = \left\langle f(t), \psi\left(\frac{t-b}{a}\right)\right\rangle = \frac{1}{\sqrt{a}}\int_{-\infty}^{\infty} f(t)\psi\left(\frac{t-b}{a}\right)dt \quad (4.19)$$

For a fixed scale, a, the CWT is the convolution of the signal $f(t)$ with the time reversed wavelet $\psi(-(t-b)/a)$.

Figure 4.4 shows a signal $f(t)$ along with the Morlet wavelet at three scales and shifts. As a increases, the wavelet stretches by a factor of two, which analyses the signal at lower frequency; as b increases, the wavelet shifts right, which analyses the signal at different localisations. The variations of a and b lead to the multiscale and localised analysis with wavelet.

In physical interpretation, the coefficients of the wavelet transform indicate the variation of energy of the signal with time and frequency. In engineering applications, the square of the coefficients of the CWT is often called a scalogram,

Figure 4.4 Morlet wavelet transform at different scales and locations

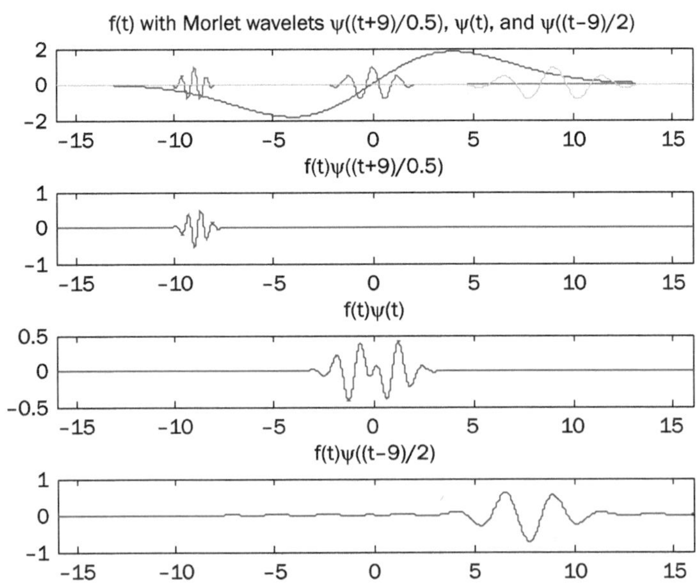

defined as Equation 4.20, which has been widely used for machinery monitoring and TCM.

$$SC_t(a,b;\psi) = \left|W_\psi f(a,b)\right|^2 \tag{4.20}$$

Comparison of time-frequency resolution of FT and WT

The time-frequency resolution of Fourier methods and wavelet analysis are illustrated in Figure 4.5. Figure 4.5a is T-F resolution for the time signal. As can be seen from the figure, it has infinite time resolution but has no frequency resolution. Figure 4.5b is the T-F resolution for FFT transformed signal. On the contrary, it has no time resolution but infinite frequency resolution. Figure 4.5c is the T-F resolution of STFT. Once the window is chosen, the resolution is fixed. Figure 4.5(d) is the T-F for wavelet analysis. The T-F boxes are adaptive. Figure 4.5a–c can be regarded as special cases of Figure 4.5d. The signal is localised

Figure 4.5 Time-frequency resolution of Fourier transform and wavelet transform

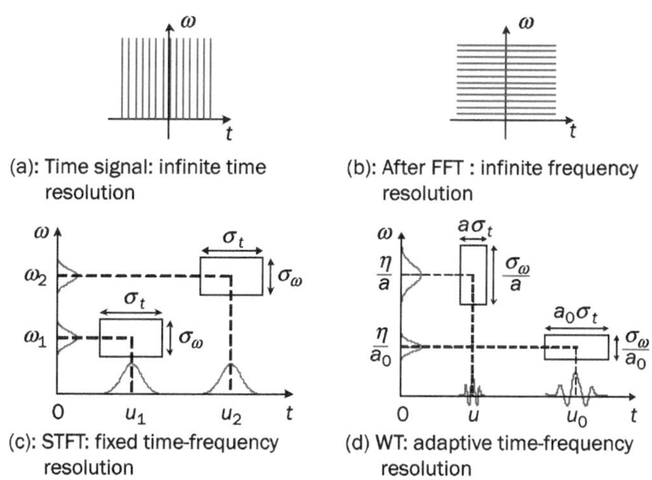

in the area with time width $\Delta_t: [u_0 - \frac{1}{2}a_0\sigma_t, u_0 + \frac{1}{2}a_0\sigma_t]$ and frequency width $\Delta_\omega: [\frac{\eta}{a_0} - \frac{\sigma_\omega}{2a_0}, \frac{\eta}{a_0} + \frac{\sigma_\omega}{2a_0}]$. The area of Heisenberg box remains to,

$$\Delta_t\Delta_\omega = (a_0\sigma_t)\times(\sigma_\omega / a_0) = \sigma_t\sigma_\omega \tag{4.21}$$

Small a: CWT resolve events closely spaced in time.
Large a: CWT resolve events closely spaced in frequency.

Compared with the STFT, whose time-frequency resolution is constant, the time-frequency resolution of the wavelet transform depends on the frequency of the signal. At high frequencies, the wavelet reaches a high time resolution but at a low frequency resolution; whereas at low frequencies, high-frequency resolution and low time resolution can be obtained. Such adaptability of time-frequency analysis reinforces the important status of the wavelet transform in the fault diagnostics field.

4.3.3 Discrete wavelet transform and its implementation

Discrete wavelet transform (DWT)

The CWT is recognised as an effective tool for both stationary and non-stationary signals. However, the CWT transforms the signal of one variable into a signal of two variables. Clearly, it involves much redundant information and is computationally very slow. To 'minimise' the transformation one can select discrete values of a and b and still have a transformation that is invertible.

The redundancy in the continuous wavelets can be fixed. First the continuous variables a and b are discretised so that the mother wavelet can only be scaled and translated in discrete steps. The discretised values of b are made dependent on a in such a manner that low frequency components are sampled less often. As a result of the discretisation we get the discretised wavelets,

$$\psi_{a,b}(t)=\frac{1}{\sqrt{a_0^j}}\psi\left(\frac{t-kb_0a_0^j}{a_0^j}\right) \tag{4.22}$$

The transform must be invertible to be practically useful. It has been proven that for any function (signal) $f \in L_2(R)$, if Equation 4.23 is satisfied, the transform is invertible,

$$A\|f\|^2 \le \sum_{j,k}\left|\langle f,\psi_{j,k}\rangle\right|^2 \le B\|f\|^2, \quad 0<A\le B<\infty \tag{4.23}$$

The wavelet set $\{\psi_{j,k}\}$ is called a frame. $A < B$ means the wavelets are redundant, and when the equality holds it is called a tight frame. The tight frame with $A = B = 1$ actually discretises the wavelets to form an orthonormal basis (orthogonal basis, with norm-2 equal to one) as in this condition, $\|f\|^2=\sum_{j,k}\left|\langle f,\psi_{j,k}\rangle\right|^2$.

The orthogonal condition is, for all values of j,k,m,n,

$$\int \psi_{j,k}(t)\psi^*_{n,m}(t)\,dt=\delta(j-n)\delta(k-m) \tag{4.24}$$

The equation holds only if $j=n$ and $k=m$. Notice that the orthogonality is not essential for discrete wavelets to work, and sometimes it is not required, as it reduces the flexibility of the wavelets.

So the key to wavelet analysis is to construct a wavelet function $\psi_{j,k}(t)$, so that the wavelets $\{2^{-j/2}\psi(2^{-j}t-k),\ i,k \in Z\}$ form an orthononal basis in $L_2(R)$. Mallat (1989) developed a framework for such a construction under the idea of multiresolution analysis (MRA) (refer to Mallat, 1989, 2008 for details). Basically, an MRA consists of a function space

$\{V_j\}_{j\in Z}$ of $L^2(R)$ satisfying some important properties, such as scaling, approximation, nesting and shifting properties. The function space has nesting property such that $V_0 \supset V_1 \ldots V_j \supset V_{j+1} \ldots$, and the function defined in the space $\{V_j\}_{j\in Z}$ with $\phi(t) \in V_j \Leftrightarrow \phi(\frac{t}{2}) \in V_{j+1}$, is called a scaling function. By defining W_k as an orthogonal complement of V_k in V_{k-1}, i.e. $V_{j-1} = V_j \oplus W_j$ and $V_k \perp W_k$. The space V_j are the sums of W_j:

$$V_0 = W_1 \oplus W_2 \oplus \ldots \oplus W_j \oplus W_{j+1} \oplus V_{j+1} \tag{4.25}$$

This construction can be geometrically viewed through Figure 4.6, the function x can be approximated by the basis ψ and ϕ, and the scaling basis ϕ could be further approximated by finer wavelet and scaling basis.

According to Mallat's multiresolution analysis (MRA), the discretising of (a, b) that preserves all information about the decomposed function cannot be coarser than the critical sampling, and any coarse discretisation will not give a unique inverse transformation. Under the critical sampling, a_0 is assigned a value of 2 and b_0 the value of 1. By discretising $(a, b) = (2^j, k*2^j)$, we have the wavelet function

$$\psi_{j,k}(t) = 2^{-j/2}\psi(2^{-j}t - k), \tag{4.26}$$

and define an accompanying scaling function as

$$\phi_{j,k}(t) = 2^{-j/2}\phi(2^{-j}t - k) \tag{4.27}$$

Figure 4.6 **Geometry of MRA**

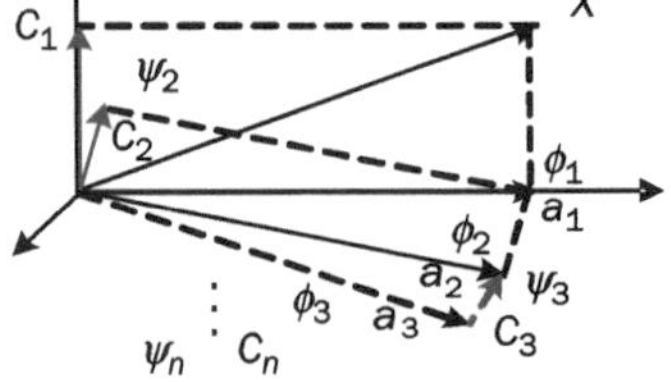

$$\begin{aligned} X &= C_1\psi_1 + a_1\phi_1 \\ &= C_1\psi_1 + C_2\psi_2 + a_2\phi_2 \\ &= C_1\psi_1 + C_2\psi_2 + C_3\psi_3 + a_3\phi_3 \\ &= C_1\psi_1 + C_2\psi_2 + C_3\psi_3 + \ldots + C_n\psi_n + a_n\phi_n \end{aligned}$$

at level of detail j and shift k.

The signal $f(t)$ is then represented by shifted and dilated versions of $\psi(t)$ and $\phi(t)$ with discrete wavelet transform (DWT),

$$f(t) = \sum_{k \in Z} c_{j_0,k} \phi_{j_0,k}(t) + \sum_{j \ge j_0} \sum_{k \in Z} d_{J,k} \psi_{J,k}(t) \tag{4.28}$$

where

$$c_{j,k} = 2^{-j/2} \int_R f(t) \phi(2^{-j} t - k) dt \tag{4.29}$$

is the scaling coefficient, which measures the local mean around the time $2^{-j}\, i$, and

$$d_{j,k} = 2^{-j/2} \int_R f(t) \psi(2^{-j} t - k) dt \tag{4.30}$$

is the wavelet coefficient, which characterises the local variation around the time $2^{-j}\, i$ and the frequency $2^{-j} f_0$.

DWT implementation with filter banks

Practically, the DWT is implemented using a filter bank structure above. The scaling function $\varphi(t)$ lies in the central space V_0, so $\varphi(t) \in V_0 \subset V_1$ since the spaces are nested. Also, the translations $\varphi(2t - k)$ of $\varphi(2t)$ form an orthonormal basis for V_1.

Since $\varphi(t) \in V_1$, we can write $\varphi(t)$ in terms of this basis, and relate it to the two-scale difference equation (Mallat, 1989):

$$\phi(t) = \sqrt{2} \sum_k h_0(k) \phi(2t - k) \tag{4.31}$$

where coefficient $h(k)$ must satisfy several conditions to make bias wavelet functions unique, orthonormal and have a certain degree of regularity. The mother wavelet is related to the scaling function via:

$$\psi(t) = \sqrt{2}\sum_{k} h_1(k)\phi(2t-k) \tag{4.32}$$

where

$$h_1(k) = (-1)^k h(1-k) \tag{4.33}$$

Note that $h_0(k)$ and $h_1(k)$ can be viewed as filter coefficients of half band lowpass and highpass filters, respectively. At this point, if valid $h_0(k)$ is available, one can obtain $h_1(k)$. Different $h(k)$ and scaling function may lead to different qualified wavelets such as Haar, Morlet and Daubechies.

Figure 4.7a illustrates the five-level MRA analysis of a cutting force signal sampled at 6,000Hz. With a five-level

Figure 4.7 MRA analysis of cutting force and their frequency bands

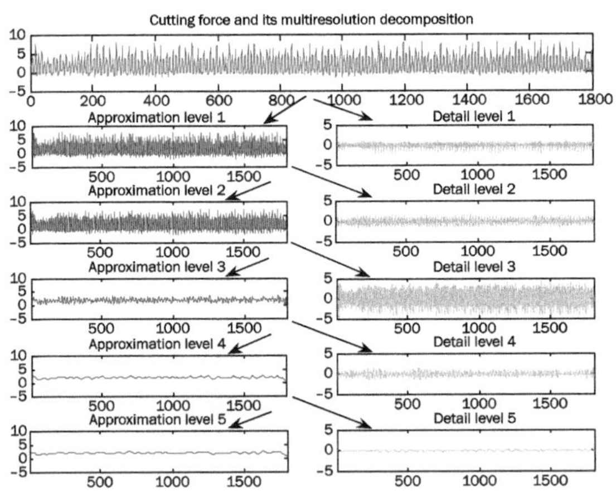

(a) Five-level MRA analysis of cutting force

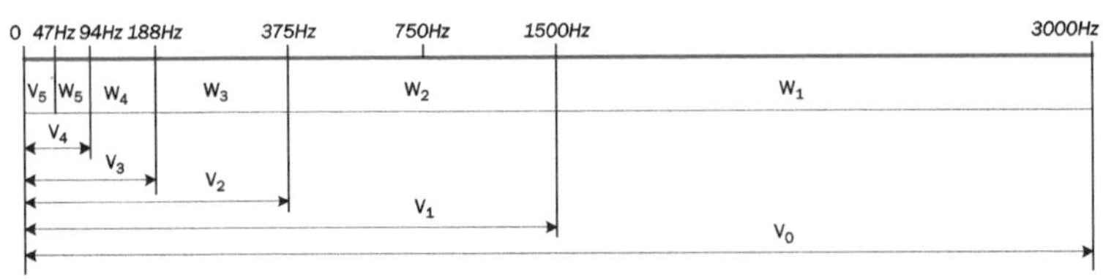

(b) Frequency band separation of five-level MRA analysis

MRA analysis, the corresponding frequency bands are separated as illustrated in Figure 4.7b. By DWT, the signal $f(t)$ is decomposed into two parts: low-pass approximation coefficients and high-pass detail coefficients. The next step then decomposes the new approximation coefficients.

4.3.4 Wavelet packets decomposition

The DWT may lead to a loss of useful information at high frequency, since successive details are no longer analysed. We need to double the sampling rate for higher frequency analysis, which, however, involves more data and computation. Wavelet packet decomposition (Wickerhauser and Coifman, 1992) is a generalisation of wavelet decomposition at higher frequencies. In the wavelet packet decomposition, each detail coefficient vector is also decomposed into two parts using the same approach as DWT in approximate coefficients. This offers the analysis at the high frequencies.

Basically, in WPD, for fixed values of j and k, $W_{j,n,k}$ analyses the fluctuations of the signal roughly around the position $2^j \cdot k$, at scale 2^j and oscillation parameter n. WPA operates by approximating a signal with scaled and translated wavelet packet functions,

$$W_{j,n,k} = 2^{-j/2} W_n(2^{-j/2} t - k) \tag{4.34}$$

The notation which characterises each wavelet packet $W_{j,n,k}$ reflects the scale 2^j and location 2^j k. The wavelet packet coefficients are then produced from the integral,

$$d_j^k = \int W_{j,n,k}(t)\, f(t)\, dt \tag{4.35}$$

It should be emphasised that Equation 4.35 allows many possible combinations of wavelet packet functions to be selected in order to optimally characterise the signal. Several criteria have been developed to select the best basis for these purposes (Wickerhauser and Coifman, 1992).

4.3.5 Some remarks on wavelet transform

Regularity

The regularity condition requires that the mother wavelet has to be locally smooth and concentrated in both the time and frequency domains. To ascertain a good local property in the frequency domain, the $|W_\psi f(a,b)|$ is required to decay rapidly with decrease of scale a. As a result this requires the first m moments of $\psi(x)$ equal to zero. These are the vanishing moments. If the Fourier transform of the wavelet is M times differentiable the wavelet has M vanishing moments.

$$\int x^m \psi(x)\,dx = 0,\ m \in [0, M] \tag{4.36}$$

This means the wavelet is 'blind' to polynomial segment with degree $\alpha \leq M-1$. Physically speaking, a higher degree of polynomial means higher frequency (faster variations of the signal); as a result, the higher moment indicates that wavelet is a high-pass filter.

Accuracy of signal approximation

It has been proved (Daubechies, 1990) that the accuracy of wavelet approximation of function (signal) $f(t)$ in the vicinity of t_0 has the following upper bound,

$$\left|\langle f, \psi_a \rangle - f(t_0)\right| \leq \frac{2C}{M!}\left(\frac{R}{a}\right)^M \tag{4.37}$$

where R is the support area, C is a constant related to f.

Two facts can be derived from this equation:

1. The accuracy of wavelet approximation $|\langle f, \psi_a \rangle - f(t_0)|$ depends on the support area R of wavelet function ψ_t and scaling parameter a. If support area R is compact (limited and close bounds) support and the scaling parameter a approach infinity, $a \to \infty$, then $|\langle f, \psi_a \rangle - f(t_0)| \to 0$.
2. In the case of the scaling parameter $a > R$, the accuracy of $|\langle f, \psi_a \rangle - f(t_0)|$ increases with the increase of vanish moment M.

As a result, from the function approximation point of view, we hope the wavelet ψ_t with smaller support R and higher vanish moment M. At the same time, with smaller R, we get a better time resolution. According to the uncertainty principle, the frequency resolution decreases.

When choosing a particular wavelet, we thus face a trade-off between the number of vanishing moments and the support size. Generally, if f has few isolated singularities and is regular between singularities, we may like to choose a wavelet with higher vanishing moments to produce a large number of small wavelet coefficients; if the density of singularities increases, it might be better to decrease the size of its support at the cost of reducing the number of vanishing moments. The Daubechies wavelet has the shortest support $\propto 2M$ for all wavelet families with M vanishing moments.

Continuous or discrete analysis

When the signal is recorded in continuous time or on a very fine time grid, both analyses are possible. Which should be used? It depends; each one has its own advantages. DWT ensures space-saving coding and is sufficient for exact reconstruction. CWT is often easier to interpret, since its redundancy tends to reinforce the traits and makes all information more visible. This is especially true of very subtle information. Thus, the CWT gains in 'readability' and in ease of interpretation, but loses in terms of saving space.

Choice of scale

A small scale value permits us to perform a local analysis; a large scale value is used for a global analysis. Combining local and global is a useful feature of the wavelet analysis. The wavelets having a compact support are used in local analysis. This is the case for Haar and Daubechies wavelets. As wavelet analysis constraint to Heisenberg uncertainty principle, it is impossible to reduce arbitrarily both time and frequency localisation. The resolution increases as the scale decreases. The greater the resolution, the smaller and finer are the details that can be accessed.

Relationship between scale and frequency

As wavelet analysis is time-scale analysis, the scale is not directly related to frequency. Abry (1997) provided a way to associate them. The idea is to associate with a given wavelet a purely periodic signal of frequency F_c. The frequency maximising the *FFT* of the wavelet modulus is called centre frequency F_c. The centre frequency captures the main wavelet oscillations. When the wavelet is dilated by a factor a, this centre frequency becomes F_c/a. We then can associate to the scale a the frequency

$$F_a = \frac{F_c}{a \cdot T_s} \tag{4.38}$$

F_a is the pseudo-frequency corresponding to the scale a, in Hz.

4.4 Sparse coding

The Shannon/Nyquist sampling theorem (see 'Linear system, sampling theorem and convolution' above) specifies that to avoid losing information when capturing a signal, one must sample at least two times faster than the signal bandwidth. In many applications, for example in TCM with acoustic emission (AE), the Nyquist rate is so high (>Mb) that too many samples result, making it impossible for real-time application. This section presents a new method to capture and represent compressible signals at a rate significantly below the Nyquist rate. This method, called sparse coding, or compressive sensing in information theory (Candès et al., 2006; Donoho, 2006), employs linear projections that can preserve the structure of the signal and reconstruct it from these projections.

The idea of sparse coding is to represent a signal with linear combination of atoms in an overcomplete dictionary, so that the representation is sparse (Lewicki and Sejnowski, 2000). Sparse coding has received a great deal of attention in recent years, from signal classification to image recognition (Zibulevsky, 2001; Mallat, 2008; Mairal et al., 2008).

In signal processing, a signal is said to be sparse if most of its entries are equal to zero. A k-sparse signal is a signal for which exactly k samples have a non-zero value. Sparse coding means representing a signal as a linear combination of a few atoms of a given (often overcomplete) dictionary (Lewicki and Sejnowski, 2000). The code has to be a fair representation of the signal while it should also be sparse. Without loss of generality, the signal y can be modelled as the linear combination of T elementary waveforms, also called *signal atoms*, such that,

$$y = \Phi x = \sum_{i=1}^{T} x_i \varphi_i \tag{4.39}$$

where x_i are called the representation coefficients of y in the *dictionary* $\mathbf{D} = (\varphi_1, \ldots, \varphi_T)$. Clearly, **x** and **y** are equivalent representations of the signal, with **x** in the time or space domain and **y** in the Ψ domain. The signal **x** is *K-sparse* if it is a linear combination of only K basis vectors; that is, only K of the y_i coefficients in (1) are non-zero and $(N - K)$ are zero. The case of interest is when $K \ll N$. The signal **x** is *compressible* if the representation (1) has just a few large coefficients and many small coefficients.

There are several overcomplete dictionaries that are built from common bases. These include local cosine dictionary, overcomplete wavelets, curvelets, and Gabor dictionaries (Mallat, 2008). These customised dictionaries are appealing because their corresponding representation can be found efficiently by leveraging their compatibility with fast transforms. Under certain circumstances, more elaborate dictionaries can be constructed by taking the union of different dictionaries. Figure 4.8 gives an example of a compressive sensing measurement process with overcomplete dictionary representation (sparse coding). The measurement vector y has 20 non-zero coefficients. The coefficient vector x is sparse with K = 10 in terms of the overcomplete dictionary

Figure 4.8 **Sparse representation of the signal vector**

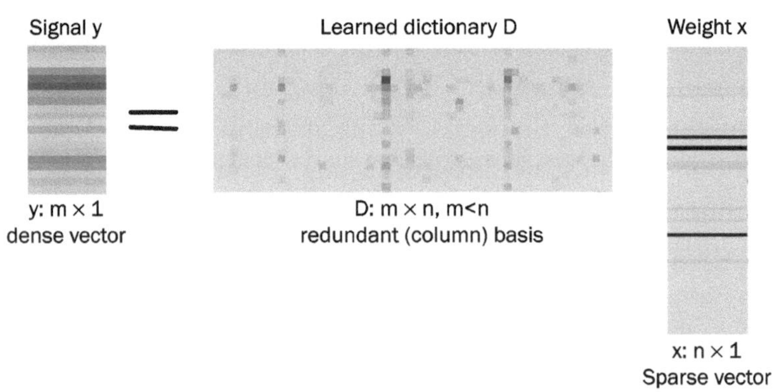

D in a higher dimension. In text categorisation, the basis elements of dictionary D allow recovering different topics and weights allow assigning each text to its corresponding category.

The key notions in the development of sparse decomposition theory are *sparsity* and *incoherence* (Candes et al., 2006). Sparsity expresses the idea that the information carried by the signal may be much smaller than its bandwidth. The incoherence means that the basis must be significantly different so that a dictionary composed by such bases has a rich vocabulary.

A common approach for sparse coding (see for example Lewicki and Sejnowski, 2000) is to use mathematical optimisation to find the best code for signal/image representation. Before coming to the optimisation function, we need to define the norms of a vector.

Definition: l_p-*norm* (Mallat, 2008)

Given a vector $x = [x_1, x_2, \ldots, x_n]^T$, the l_p-*norm of* x is defined by

$$\|x\|_p = \left[\sum_{i=1}^{n} |x_i|^p\right]^{1/p} \tag{4.40}$$

In particular, $\|x\|_1 = \sum_{i=1}^{n} |x_i|, \|x\|_2 = \left[\sum_{i=1}^{n} |x_i|^2\right]^{1/2}, \|x\|_\infty = \max_{1 \le i \le n}\{|x_i|\}$.

In the literature, the sparsity of a signal (vector) x is often related to the so-called l_0 norm defined by

$$\|x\|_0 = \sum_{i=1}^{n} |x_i|^0 = \text{number of nonzeros in } x \tag{4.41}$$

Now, we could come to the mathematical representation of SC. Given a signal $y \in R^n$ and a dictionary $D \in R^{n \times k}$, the sparse representation problem can be stated as

$$\min_x \|x\|_0 \quad \text{s.t. } y = Dx, \tag{4.42}$$

where $\|x\|_0$ is the 'l_0-norm' of the coefficient vector $x \in R^k$, the number of non-zero elements. This form is known as compressive sensing in information theory. This problem is NP-hard, and is commonly approximated by substituting the l_0-norm in Equation 4.42 with l_1-norm. An efficient approach to this is to solve it as a Lasso-type problem (Tibshirani, 1996),

$$\min_{\alpha} \|y\text{-}Dx\|_2^2 + \lambda \|x\|_1 \ , \qquad (4.43)$$

where λ is a parameter that balances the trade-off between reconstruction error and sparsity.

It is a known fact that in general the l_1 constraint induces sparse solutions for the coefficient vectors x. Furthermore, this is a convex problem that can be solved very efficiently using for example the LARS-Lasso algorithm (Efron et al., 2004). This alternative has also been shown to be more stable than the l_0 approach.

Figure 4.9 shows an example of a signal that is dense in time domain but after Fourier transform, it is nearly sparse in frequency domain (0–500Hz) as most coefficients are close to zero. In this condition, even with much lower frequency component of the signal (0–60Hz), it could be

Figure 4.9 **Signal reconstructed below Nyquist frequency**

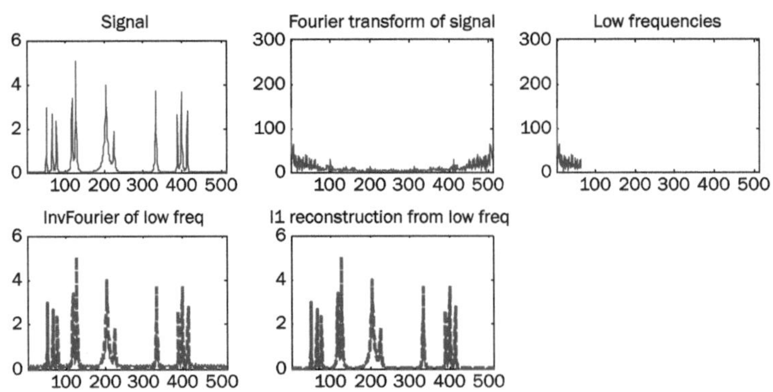

nearly recovered with the l_1 approach. This is promising in TCM for example with AE and force processing. We can sample the signal in much lower rate that could be analysed in real time, which had been far from implementation.

4.5 Applications

We first present a review of the time-frequency analysis of machining signals, signal de-noising and feature extraction of wavelet applications for TCM. These three approaches are studied in most of the literature. Two new approaches, called singular detection and sparse signal de-noising, are introduced with real experiments and discussed for TCM. The applications of WT for feature extraction and dimension reduction have been of general interest and widely studied, and the case study is not given here as a result. Interested readers may like to refer to Zhu et al. (2009b).

4.5.1 Sparse coding for de-noising in the STFT domain

Signal noise always exists in machining. De-noising is a common practical problem in signal processing. Wavelet de-noising has been employed in TCM in some studies. Kwak and Ha (2004) described the use of the grinding force signal with noise reduction to detect the dressing time based on DWT. As a result of de-noising, the grinding force signal was successfully used to detect the need for dressing. The wavelet de-noising method was found to be more effective than the FFT filtering technique. Li and Guan (2004) proposed a wavelet-based de-noising to extract marked features from the feed-motor current signals to indicate the minor cutting-edge fracture. It must be mentioned that the existing wavelet de-noising methods reported in the literature rely heavily on white Gaussian noise and relative energy levels of wavelet coefficients. The wavelet de-noising methods encounter problems in de-noising non-Gaussian noise, which is discussed in Zhu et al. (2008a, 2009a).

Sparse decomposition techniques for signals and images underwent considerable development during the flourishing of wavelet-based compression and de-noising methods (Zibulevsky and Pearlmutter, 2001; Mallat, 2008). Their main impact is that they provide a relatively simple framework for separating a number of sources exceeding the number of observed mixtures. Also they greatly improve quality of separation in the case of square mixing matrix.

There exist several well-known orthonormal bases, such as the Fourier basis, wavelet basis, or the discrete cosine transform, that are well used throughout the signal processing literature. Many of these bases do not require matrix multiplication because there are fast transforms such as the fast-Fourier transform (FFT) that are more computationally efficient. These transforms and their derivatives also result in sparse representation for a certain class of signals. For example, the short-time Fourier transform (STFT) generates sparse time-frequency representation of force signals. In Figure 4.10, we illustrate a force signal in the time domain and its corresponding representation using a STFT. As we can see, the signal has non-zero coefficients for almost every time point in the time domain, but in the time-frequency domain, only some time-frequency blocks have a relevant coefficient.

In this study, the force de-noising is solved as a blind source separation (BSS) problem. Blind source separation (Comon, 1994) is a form of signal separation technique which blindly estimates individual source components from their received mixtures at sensors. In BSS, the m observed signals $x_i(k)$ $(i = 1,2,\ldots m)$ are linear mixed combinations of n

Figure 4.10 The force's sparse representation in STFT domain

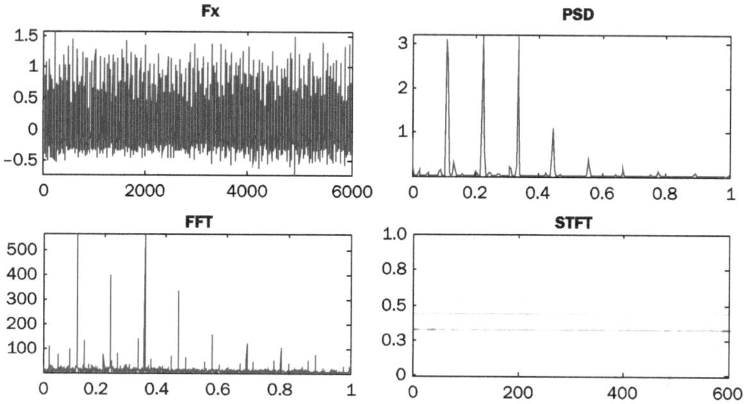

(typically $m \geq n$) unknown mutually statistically independent source signals $s_i(k)$ $(i = 1,2,\ldots n)$, and noise $v_i(k)$ $(i = 1,2,\ldots m)$. This can be expressed as

$$x_i(k)\ (i=1,2,\ldots m)=\sum_{j=1}^{n} h_{ij}s_j(k)+v_i(k) \tag{4.44}$$

The aim is to find some methods to un-mix this process and find the original sources. Assume that there is a matrix W and

$$Y(t)=W\mathbf{x}(t)=W[H\mathbf{s}(t)]=(WH)\mathbf{s}(t) \tag{4.45}$$

If $WH = I$, i.e., $W = H^{-1}$, then the $Y(t)$ is regarded as an estimation of $\mathrm{s}(t)$. In this way, if we can find an un-mixing matrix W, we will recover the sources $\mathbf{s}(t)$.

In order to perform the separation, we perform STFT over the forces. It is expected that the sources will be well separated over the Fourier domain because the sources are sparse after a STFT. The signal in time domain $x_i(n)$ at position k was transformed with STFT into the frequency domain,

$$\hat{X}_i(t,f)=\sum_{\tau=-K/2}^{K/2-1} x_j(\tau+t)win(\tau)e^{-2j\pi f\tau} \tag{4.46}$$

Where $f = 0,(1/K)f_s,...,(K-1/K)f_s$, f_s is sampling frequency and $win(\tau)$ is the window function.

By carefully choosing the window, this transform corresponds to the decomposition of the signal in a redundant tight frame. The redundancy corresponds to the overlap of the windows, and the tight frame corresponds to the fact that the pseudo-inverse is simply the transposed of the transform (it means that the same window can be used for synthesis with a simple summation of the reconstructed signal over each window).

Then the mixture Y in the time domain could be represented in the frequency domain as

$$\hat{X}(t,f)= \hat{A}(f)\hat{S}_i(t,f) \tag{4.47}$$

$\hat{A}(f)$ is a $M \times N$ mixture matrix in the frequency domain, $\hat{X}(t, f)$ and $\hat{S}_i(t, f)$ are the observation and sources in the frequency respectively. The unmixing model is then expressed in the frequency domain as

$$\hat{Y}(t,f) = \hat{W}(f)\hat{X}(t,f) \tag{4.48}$$

where $\hat{W}(f)$ is the unmixing matrix in the frequency domain.

The inverse transform is then applied to this formula to reconstruct the signal in the time domain,

$$\hat{y}(t)= \hat{w}\hat{x}(t) \tag{4.49}$$

Figure 4.11 shows a typical sample of the cutting force and noise from the micro-milling experiment. As can be seen, the SNR is very low and the absolute value of the noise is nearly one-third that of the force. The noise statistics are computed as: mean = –0.1093, variance = 0.2631, skew = 0.0004, kurtosis = 1.2621. It is super-Gaussian, with a longer tail than Gaussian distribution and is well fitted with Laplacian distribution (see Figure 4.11b). The power spectrum density (PSD) and autocorrelation

Figure 4.11 Illustrative of the statistics of noise: a) noise signal, b) noise distribution compared to Gaussian distribution, c) in frequency domain, d) the autocorrelation coefficients

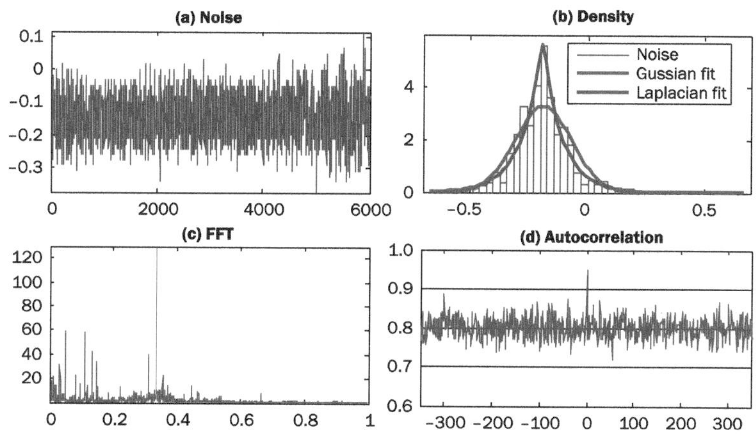

show that the noise is not white. The PSD harmonics peak at the force's characteristic frequencies (*nFr*) and the autocorrelation is very high. Typically it resembles a component of cutting force.

Figure 4.12 shows the three force signals and their corresponding power spectrum at medium flank wear in precision milling. As can be seen from Figure 4.12, the low-frequency components have relatively larger peaks

Figure 4.12 The sensor output and their corresponding power spectrum

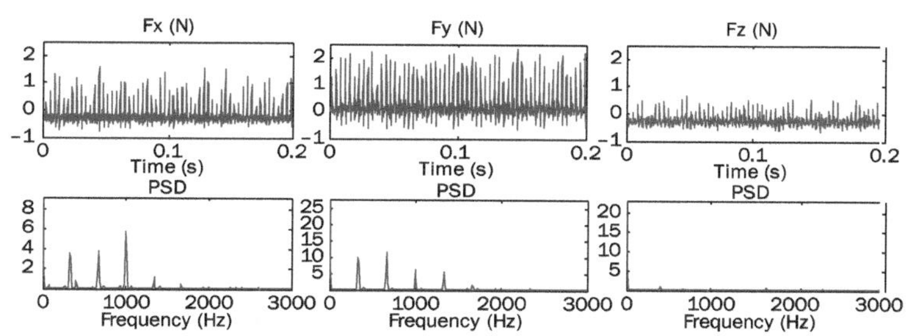

Figure 4.13 Cutting forces reconstructed

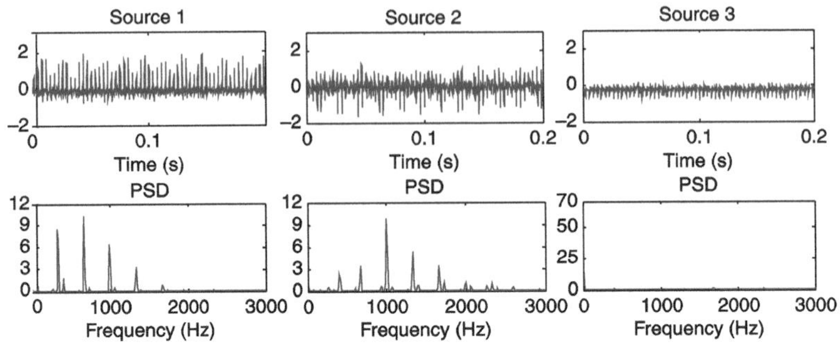

and, as a result, the desired harmonics *nFr* are immersed. This phenomenon is even worse when the tool flank wear value is higher. The reconstructed sources and their corresponding power spectrum are shown in Figure 4.13. It can be seen that reconstructed force peaks at the first five harmonics, while the machining noise peaks at low frequency and a wide range of harmonics, and the environmental noise peaks at very low frequency. The reconstructed (de-noised) force component could then be integrated to provide useful and clear trend analysis in case of tool wear increases (see Zhu et al., 2011b, for details).

For a comparison, we also show the residue of wavelet de-noising with Donoho's soft thresholding (Donoho, 1995) in Figure 4.14. As can be seen, the residue of wavelet de-noising is typically a Gaussian noise.

4.5.2 Singularity analysis for tool wear detection

Singularities can be thought of as either an abrupt change or 'impulse' in a signal, or the sudden shift of the signal's mean value to a different level. The good time-frequency localisation property provides wavelet in singularity analysis (Mallat and

Figure 4.14 **Residue of wavelet de-noised force**

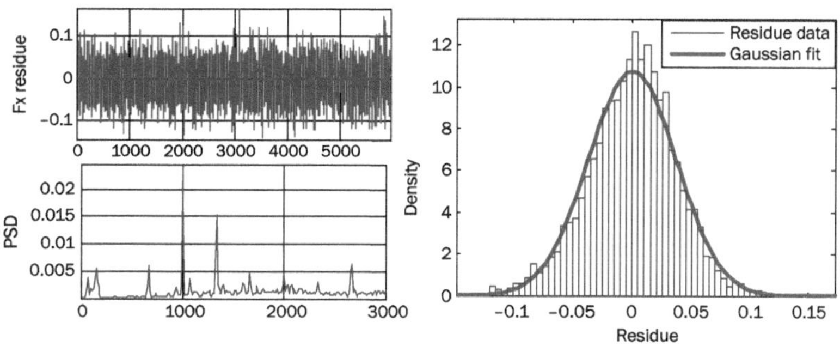

Hwang, 1992). The general idea used to detect a function's abrupt changes through wavelet coefficient regularity at a certain point and the rate of decay of the coefficients near this point. Beyond the application in image edge detection, the idea of singularity analysis with wavelet is implemented in machinery condition monitoring, which is similar to the TCM in machining. Sun et al. (2005) proposed singularity analysis for bearing defect diagnosis with the wavelet transform modulus maxima of vibration. Loutridis and Trochidis (2004) also employed the Holder exponent to investigate gear faults. To the best of our knowledge, there has been little research on TCM with wavelet singular detection (Zhu et al., 2011a). Because it is robust and stable in capturing signal changes, it can be expected that these singularity-based features will provide a valuable study on TCM.

The singularity of the signal is measured by the Lipschitz exponent (LE). Assume the signal $f(t)$ has the following property in the vicinity of t_0,

$$\left|f(t_0+h)-P_n(h)\right| \le A\left|h\right|^{\alpha}, n<\alpha<n+1 \tag{4.50}$$

where $P_n(t)$ is an n-order polynomial and pass the point $f(t_0)$, and h is small enough, the Lipschitz exponent (LE) of $f(t)$ at t_0 is the superemum of the α for which Equation 4.50 holds.

It follows directly that if α determines $f(t)$'s continuously differentiable times at t_0, the Lipschitz exponent (LE) α gives an indication of how regular the function $f(t)$ is at t_0. The higher the α, the more regular the function $f(t)$ is.

It is geometrically illustrated in Figure 4.15.

For example (with reference to Figure 4.16) white noise HE = –0.5-ε (ε >0), ramp signal is piece-wise linear, 1-order differentiable, HE = 1; the first order derivative of a ramp signal is step function, HE = 1–1 = 0; the second order derivative of ramp signal is impulse signal with HE = 1–2 = –1. For the real cutting force, as illustrated in Figure 4.17, it is typical the weighted combinations of all these four typical signals.

Mallat and Hwang (1992) have shown that the local regularity of a function (signal) is related to the propagation

Figure 4.15 **Geometry of wavelet singular detection**

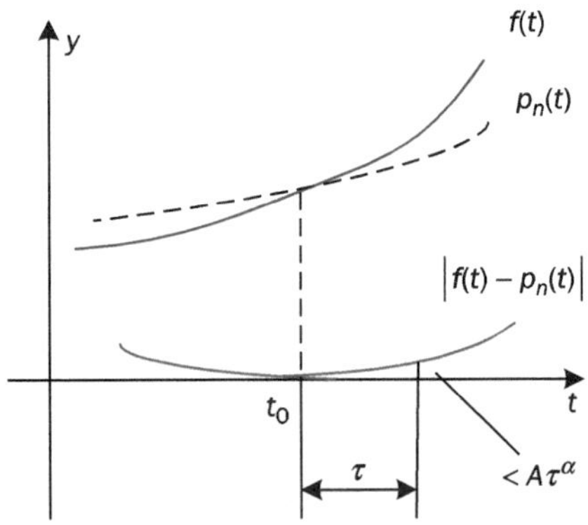

Figure 4.16 **HE value of typical signal**

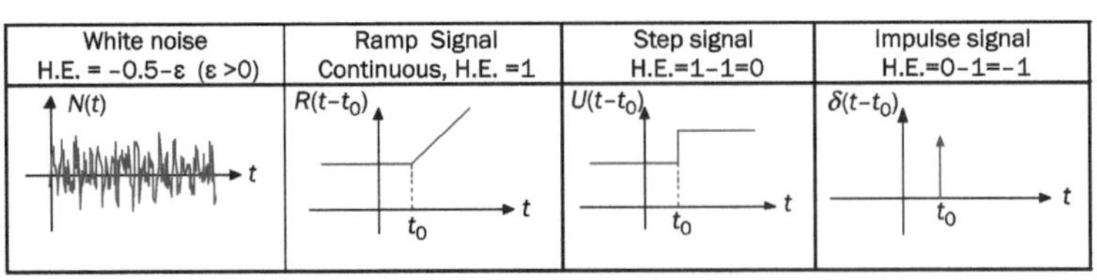

across scale of its wavelet maxima and the decay of the wavelet transform modulus in the time-scale plane. The definition of the maxima along scale:

$$\frac{dWf(s,t)}{dt}=0 \tag{4.51}$$

The modulus maxima line consists of the points that are local maxima in the time-scale plane. Along the modulus maxima line the wavelet coefficients have the scaling behaviours as follows:

$$\log_2\left|WT_{2^j}f(t)\right|\le \log_2 A+j\alpha \tag{4.52}$$

This function connects the wavelet scale j and the Holder exponent α. The A and α can be computed by setting the equality in Equation 4.52.

Figure 4.17 demonstrates an example of modulus maxima for detection of milling tool at different wear states. Figure 4.17a shows a cutting force signal from a fresh tool and severely worn tool with their wavelet modulus maxima. Lipschitz exponent (LE) extracted from the modulus maxima line (Figure 4.17b). Though they do not change much,

Figure 4.17 Modulus maxima for slight wear tool and severe wear tool

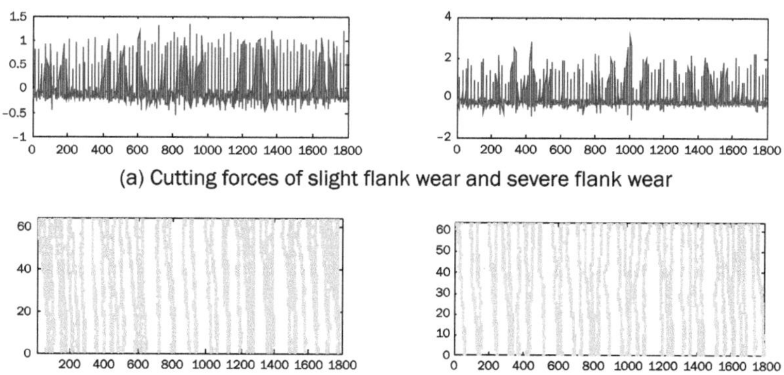

(a) Cutting forces of slight flank wear and severe flank wear

(b) Modulus maxima for slight wear tool and severe wear tool

(Continued)

Figure 4.17 *Continued*

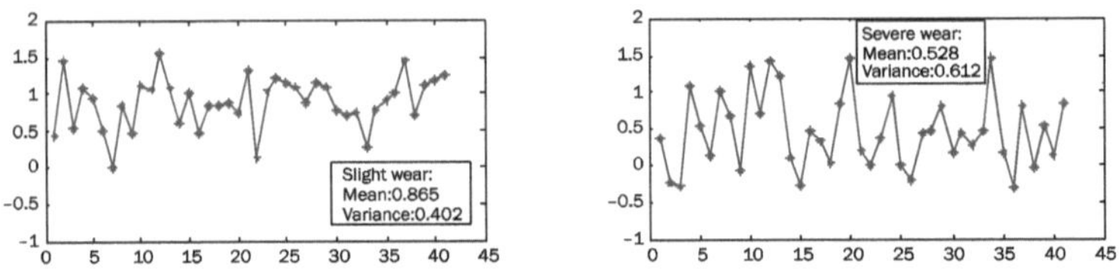

(c) Lipschitz exponent of slight wear tool and severe wear tool

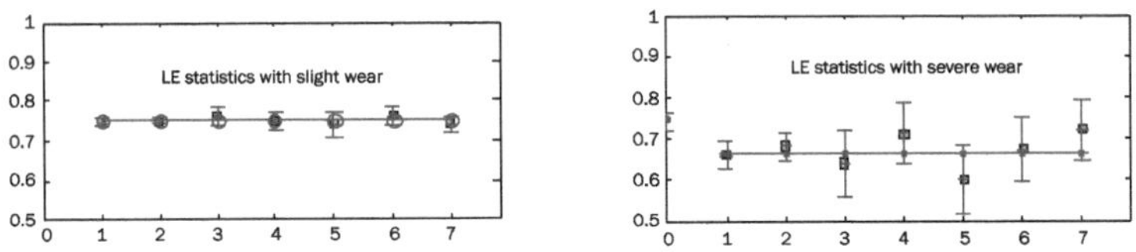

(d) Boxplot of more tests of Lipschitz exponent

Lipschitz exponents offer a stable tool to detect the degrees of tool wear level as shown. More tests and their statistics of LE from two tool states are illustrated in Figure 4.17c. It is observed that with the fault severity increasing, the milling force signals' singularities and singularity ranges will increase as well, and therefore we can evaluate the fault severity through measuring the force signals' singularities and singularity ranges.

4.6 Conclusions

In this chapter, based on the signal theory, wavelet theory and sparse decompositions are introduced with their applications to TCM. Because of the non-stationary property of the machining process, wavelet analysis and sparse decomposition are more effective thanks to Fourier methods in TCM signal analysis. Real case studies with wavelet singularity analysis and sparse decomposition for de-noising are also presented and discussed with their suitability and effectiveness. Based on the good local, sparsity and multiscale properties of these two methods, they have promising prospects in applications to TCM.

4.7 References

Abry, P. (1997) Ondelettes et turbulence. Multirésolutions, algorithmes de décomposition, invariance d'échelles, Diderot Editeur, Paris.

Bruckstein, A.M., Donoho, D.L. and Elad, M. (2009) From sparse solutions of systems of equations to sparse modeling of signals and images, *SIAM Review*, 51(1): 34–81.

Byrne, G., Dornfeld, D., Inasaki, I., Konig, W. and Teti, R. (1995) Tool condition monitoring (TCM) – the status of research and industrial application, *Annals of CIRP*, 44(2): 541–67.

Byrne, G., Dornfeld, D. and Denkena, B., (2003) Advancing cutting technology, *Annals of CIRP Keynotes.*

Candès, E., Romberg, J. and Tao, T. (2006) Robust uncertainty principles: Exact signal reconstruction from highly incomplete frequency information, *IEEE Trans. Inform. Theory*, 52(2): 489–509.

Cohen, L. (1989) Time–frequency distribution – A review, *Proceedings of the IEEE*, 77: 941–81.

Comon, P. (1994) Independent Component Analysis, a new concept? *Signal Processing*, 36(3): 287–314.

Daubechies, I. (1990) The wavelet transformation, time-frequency localization and signal analysis, *IEEE Trans. Inform. Theory*, 36: 961–1005.

Donoho, D.L. (1995) De-noising by soft-thresholding, *IEEE Trans Inform. Theory*, 41: 613–27.

Donoho, D.L. (2006) Compressed sensing, *IEEE Trans. Inform. Theory*, 52(4): 1289–306.

Dowling, M.J. (1993) Application of non-stationary analysis to machinery monitoring, ICASSP-93. *IEEE International Conference on Acoustics, Speech, and Signal Processing*, 1: 59–62.

Efron, B., Johnstone, I., Hastie, T. and Tibshirani, R. (2004) Least angle regression, *Annals of Statistics*, 32(2): 407–99.

Elbestawi, M.A., Papazafiriou, T.A. and Du, R.X. (1991) In-process monitoring of tool wear in milling using cutting force signature, *International Journal of Machine Tools & Manufacture*, 31(1): 55–73.

Franks, L.E. (1969) *Signal theory*. New York: Prentice-Hall.

Gabor, D. (1946) Theory of communication, *J. Inst. Elec. Eng.*, 93(III): 429–57.

Gong, W., Obikawa, T. and Shirakashi, T. (1997) Monitoring of tool wear states in turning based on wavelet analysis, *JSME Int. J. (Series C)*, 40: 447–53.

Kwak, J.S.(2006) Application of wavelet transform technique to detect tool failure in turning operations, *International Journal of Advanced Manufacturing Technology*, 28(11–12): 1078–83.

Kwak, J.S. and Ha, M.K. (2004) Detection of dressing time using the grinding force signal based on the discrete wavelet decomposition, *International Journal of Advanced Manufacturing Technology*, 23(1–2): 87–92.

Lee, B.Y. and Tarng, Y.S. (1999) Application of the discrete wavelet transform to the monitoring of tool failure in end milling using the spindle motor current, *International Journal of Advanced Manufacturing Technology*, 15: 238–43.

Lewicki, M.S. and Sejnowski, T.J. (2000) Learning overcomplete representations, *Neural Computation*, 12(2): 337–65.

Li, X. and Guan, X.P. (2004) Time-frequency-analysis-based minor cutting edge fracture detection during end milling, *Mechanical Systems and Signal Processing*, 18: 1485–96.

Liang, S.Y. and Dornfeld, D.A. (1989) Detection of cutting tool wear using time series modeling of acoustic emission signals, *Transactions of the ASME, Journal of Engineering for Industry*, 111(3): 199–205.

Liang, S.Y., Rogelio, L.H. and Robert, G.L. (2004) Machining process monitoring and control: The state-of-the-art, *Journal of Manufacturing Science and Engineering*, 126: 297–310.

Loutridis, S. and Trochidis, A. (2004) Classification of gear faults using holder exponents, *Mechanical Systems and Signal Processing*, 18: 1009–30.

Mairal, J., Bach, F., Ponce, J., Sapiro, G. and Zisserman, A. (2008) Discriminative learned dictionaries for local image analysis. In: *Proc. CVPR*.

Mallat, S.G. (1989) A theory of multiresolution signal decomposition: The wavelet representation, *IEEE Trans. on Pattern and Machine Intelligence*, 11(7): 674–93.

Mallat, S.G. (2008) *A Wavelet Tour of Signal Processing: the Sparse Way*, 3rd edition. New York: Academic Press.

Mallat, S.G. and Hwang, W.L. (1992) Singularity detection and processing with wavelets, *IEEE Transactions on Information Theory*, 38: 617–43.

Moon, T.K. and Stirling, W.C. (2000) *Mathematical Methods and Algorithms for Signal Processing*. New York: Prentice Hall.

Mori, K., Kasashima, N., Fu, J.C. and Muto, K. (1999) Prediction of small drill bit breakage by wavelet transforms and linear discriminant functions, *International Journal of Machine Tools & Manufacture*, 39: 1471–84.

Shannon, C.E. (1949) Communication in the presence of noise, *Proc. Institute of Radio Engineers*, 37(1): 10–21.

Sun, Q., Tang, Y., Lu, W.Y. and Ji, Y. (2005) Feature extraction with discrete wavelet transform for drill wear monitoring, *Journal of Vibration and Control*, 11(11): 1375–96.

Tansel, I.N., Rodriguez, O., Trujillo, M., Paz, E. and Li, W. (1998) Micro-end-milling – I. wear and breakage, *International Journal of Machine Tools & Manufacture*, 38: 1419–36.

Tarng, Y.S. and Lee, B.Y. (1993) A sensor for the detection of tool breakage in NC milling, *Journal of Materials Processing Technology*, 36: 259–72.

Tibshirani, R. (1996) Regression shrinkage and selection via the lasso, *J. Royal. Statist. Soc B.*, 58(1): 267–88.

Wang, W.H., Hong, G.S., Wong, Y.S. and Zhu, K.P. (2007) Sensor fusion for on-line tool condition monitoring in milling, *International Journal of Production Research*, 45(21): 5095–116.

Wickerhauser, M.V. and Coifman, R.R. (1992) Entropy based methods for best basis selection, *IEEE Trans. Inf. Th.*, 38(2): 719–46.

Wu, Y. and Du, R. (1996) Feature extraction and assessment using wavelet packets for monitoring of machining process, *Mechanical Systems and Signal Processing*, 10(1): 29–53.

Wu, Y., Escande, P. and Du, R. (2001) A new method for real-time tool condition monitoring in transfer machining stations, *ASME Journal of Manufacturing Science and Engineering*, 123: 339–47.

Yesilyurt. I. (2006) End mill breakage detection using mean frequency analysis of scalogram, *International Journal of Machine Tools & Manufacture*, 46(3–4): 450–8.

Yoon, M.C. and Chin, D.H. (2005) Cutting force monitoring in the end milling operation for chatter detection, *IMechE 2005 Proc. IMechE*, 219, Part B: *J. Engineering Manufacture*: 455–65.

Zhu, K.P., Hong, G.S., Wong, Y.S. and Wang, W.H. (2008a) Cutting force denoising in micro-milling tool condition monitoring, *International Journal of Production Research*, 46(16): 4391–408.

Zhu, K.P., Hong, G.S. and Wong, Y.S. (2008b) Discriminate feature selection for hidden Markov models in micro-milling tool wear classification, *Machining Science and Technology*, 12(3): 348–69.

Zhu, K.P., Wong, Y.S. and Hong, G.S. (2009a) Multi-category micro-milling tool wear classification with continuous hidden Markov models, *Mechanical System and Signal Processing*, 23: 547–60.

Zhu, K.P., Wong, Y.S. and Hong, G.S. (2009b) Wavelet analysis of sensor signals for tool condition monitoring: A review and some new results, *International Journal of Machine Tools & Manufacture*, 49(4): 537–53.

Zhu, K.P., Hong, G.S. and Wong, Y.S. (2011a) Multi-scale analysis of cutting forces for micro-milling tool wear monitoring, *IEEE Transactions on Industrial Electronics*, 58(6): 2512–21.

Zhu, K.P., Wong, Y.S. and Hong, G.S. (2011b) A noise separation approach for micro-milling tool condition monitoring, *IEEE Transactions on Industrial Informatics*, Submitted.

Zibulevsky, M. and Pearlmutter, B.A. (2001) Blind source separation by sparse decomposition in a signal dictionary, *Neural Computation*, 13(4): 863–82.

5

ANN modelling of fractal dimension in machining

Prasanta Sahoo and Tapan Kr. Barman

Abstract: This chapter deals with fractal dimension modelling in machining operations. The study considers the fractal dimension to describe surface roughness. Four different machining operations, including CNC end milling, CNC turning, cylindrical grinding and EDM are carried out on mild steel (AISI 1040) work-pieces. The surface roughness data are used to develop models for predicting fractal dimension using artificial neural network (ANN). From the results of all machining operations, it is seen that developed models can predict fractal dimension very accurately.

Key words: fractal dimension; surface roughness; milling; turning; grinding; EDM; ANN.

5.1 Introduction

Surfaces are irregular though they may look very smooth. When the surfaces are magnified, the irregularities become prominent. This is true for machining surfaces also. In a material removal process such as machining, unwanted material is removed and altered surface topography is obtained. The surface generated consists of inherent irregularities left by tools, which are commonly defined as surface roughness. Such a surface is composed of a large number of length scales of superimposed roughness that are generally characterised by the standard deviation of surface peaks. Three statistical characteristics are generally used to describe the structure of machined surface topography: texture, waviness and roughness. The texture determines the anisotropic property of the surface. The waviness reflects the reference profile (or surface). The surface roughness is formed by the micro deformation during the machining process.

Surface roughness plays an important role in mechanical properties such as fatigue behaviour, corrosion resistance, creep life, etc. It also affects other functional attributes of machine components such as friction, wear, light reflection, heat transmission, lubrication, electrical conductivity, etc. Surface roughness may depend on various factors such as machining parameters, work-piece materials, cutting tool properties, cutting phenomenon, etc. In a review article, Benardos and Vosniakos (2003) have presented a fishbone diagram (Figure 5.1) with parameters that affect surface roughness. It is clear that surface roughness of a machined work-piece depends on various factors. Many researchers have attempted to model surface roughness but the developed models are far from complete as it is not possible to consider all the controlling factors in a particular study. So, researchers are always alert to modelling surface roughness in a better way so that surface roughness modelling can be done more accurately.

Surface roughness is generally expressed by three types of conventional roughness parameters, namely amplitude parameters, spacing parameters and hybrid parameters. Amplitude parameters are the vertical characteristics of

Figure 5.1 Fishbone diagram showing parameters affecting surface roughness (Benardos and Vosniakos, 2003)

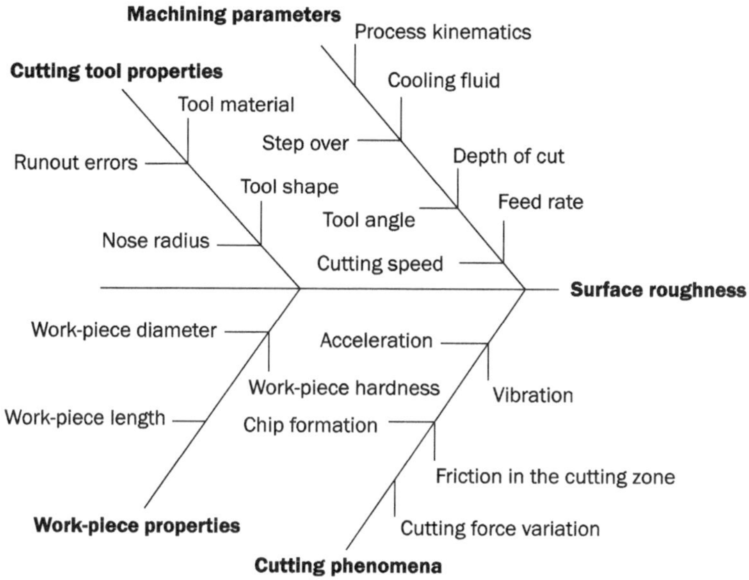

surface deviation while spacing parameters are the horizontal characteristics of surface deviation. Hybrid parameters are the combination of horizontal and vertical surface deviation. Most commonly used roughness parameters are centre line average value (Ra), root mean square value (Rq), mean line peak spacing (Rsm), etc.

Conventionally, the deviation of a surface from its mean plane is assumed to be a random process for which statistical parameters such as the variances of the height, the slope and curvature are used for characterisation (Nayak, 1971). However, it has been found that the variances of slope and curvature depend strongly on the resolution of the roughness-measuring instrument or any other form of filter and are hence not unique (Thomas, 1982; Bhushan et al., 1988; Majumdar and Tien, 1990). It is also well known that surface topography is a non-stationary random process for which the variance of the height distribution is related to the length of the sample (Sayles and Thomas, 1978). Consequently, instruments with different resolutions and scan lengths yield different values of these statistical parameters for the same surface. The conventional methods of characterisation are therefore fraught with inconsistencies which give rise to the term 'parameter rash' (Whitehouse, 1982) commonly used in contemporary literature. The underlying problem with the conventional methods is that although rough surfaces contain roughness at a large number of length scales, the characterisation parameters depend only on a few particular length scales, such as the instrument resolution or the sample length. A logical solution to this problem is to use scale-invariant parameters to characterise rough surfaces. In this context, to describe surface roughness, the concept of fractals is applied. The concept is based on the self-affinity and self-similarity of surfaces at different scales. Fractals retain all the structural information and are characterised by a single descriptor, the fractal dimension, D. Fractal dimension is an intrinsic property of the surface and independent of the filter processing. Roughness measurements on a variety of surfaces show that the power spectra of the surface profiles follow power laws. This suggests that when a surface is magnified appropriately, the magnified image looks very similar to the original surface. This property can be mathematically described by the concepts of self-similarity and self-affinity. The fractal dimension, which forms the essence of fractal geometry, is both scale-invariant and is closely linked to the concepts of self-similarity and self-affinity (Mandelbrot, 1982). It is therefore essential to use fractal dimension to characterise rough surfaces and provide the geometric structure at all length scales (Bigerelle et al., 2005). The possible application of fractal geometry to tribology was explored (Ling, 1990). The influence of processing techniques on the fractal or non-fractal structure was also examined (Majumdar and Bhushan, 1990).

In a material removal process, mechanical intervention happens over length scales which extend from atomic dimensions to centimetres. The machine vibration, clearances and tolerances affect the outcome of the process at the largest of length scales (above 10^{-3}m). The tool form, feed rate, tool radius in the case of single point cutting (Venkatesh et al., 1998) and grit size in multiple point cutting (Venkatesh et al., 1999) affect the process outcome at the intermediate length scales (10^{-6} to 10^{-3}m). The roughness of the tool or details of the grit surfaces influence the final topography of the generated surface at the lowest length scales (10^{-9} to 10^{-6}m). It has been shown that surfaces formed by electric discharge machining, milling, cutting or grinding and worn surfaces (Brown and Savary, 1991; Tricot et al., 1994; Hasegawa et al., 1996; He and Zhu, 1997; Ge and Chen, 1999; Zhang et al., 2001; Jiang et al., 2001; Zhu et al., 2003; Jahn and Truckenbrodt, 2004; Kang et al, 2005; Han et al, 2005) have fractal structures, and fractal parameters can reflect the intrinsic properties of surfaces to overcome the disadvantages of conventional roughness parameters. The present study attempts to characterise the roughness of machined surfaces in different machining processes using fractal dimension as the roughness parameter.

5.2 Basic considerations

5.2.1 Surface metrology

Surface texture is a complex condition resulting from a combination of roughness (nano and micro-roughness), waviness (macro-roughness), lay and flaw. Figure 5.2 shows a display of surface texture with unidirectional lay. Roughness is produced by fluctuations of short wavelengths characterised by asperities (local maxima) and valleys (local minima) of varying amplitudes and spacing. This occurs due to the mechanism of the material removal process. Waviness is the surface irregularities of longer wavelengths and may result

Figure 5.2 Display of surface texture

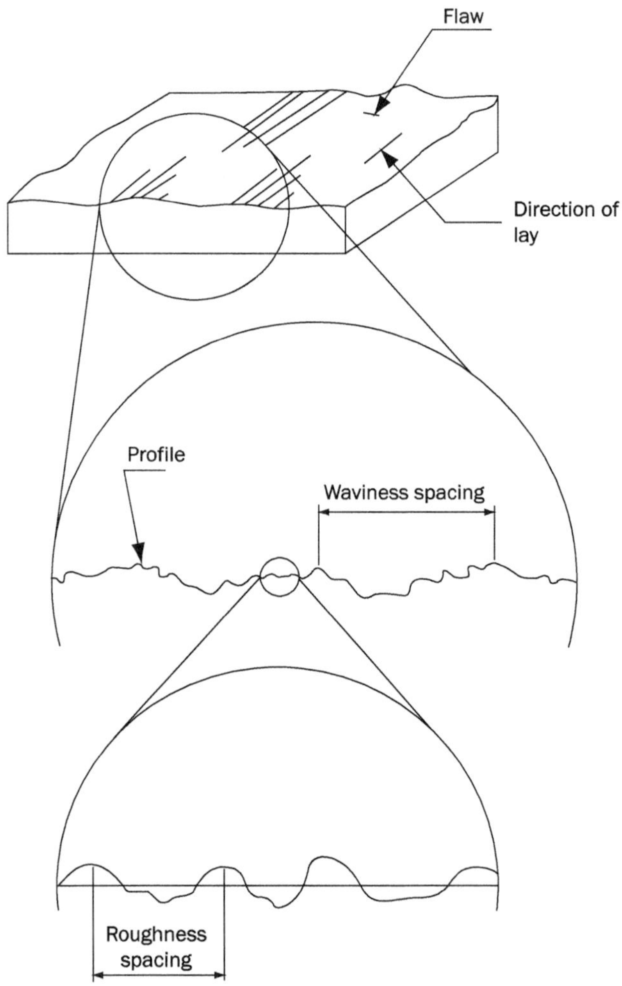

from such factors as machine or work-piece deflections, vibration, chatter, heat treatment or warping strains. Lay is the principal direction of the predominant surface pattern, usually determined by the production process. Flaws are unexpected and unintentional interruptions in the texture. Apart from these, the surface may contain large deviations from nominal shape of very large wavelength, which is known as error of form. These are not considered as part of surface texture.

Any engineering surface comprises a vast number of peaks and valleys and it is not possible to measure the height and location of each of the peaks. So measurement of a surface is carried out on a sampling length where it is assumed that the surface outside and inside the sampling length is statistically similar. In order to determine the numerical assessment of a sample's surface texture, three characteristic lengths are associated with the profile (ISO 4287:1997), namely sampling length, evaluation or assessment or cut-off length and traverse length. The sampling length is the length over which the parameter to be measured will have statistical significance. Cut-off length is the length of the surface over which the measurement is made. This length may include several sampling lengths – typically five times. The measurement is the integration of the individual sampling lengths. The total length of the surface traversed by the stylus in making a measurement is called the traverse length. It will normally be greater than the evaluation length, due to the necessity of allowing run-up and over-travel at each end of the evaluation length to ensure that any mechanical and electrical transients are excluded from the measurement.

There are several methods to study the surface topography which have been developed over the years. The most common method of studying surface texture is the surface profilometer (Figure 5.3). In this method, a fine, very lightly loaded, stylus is travelled smoothly at a constant speed across the surface

Figure 5.3 **Component parts of a typical stylus surface-measuring instrument**

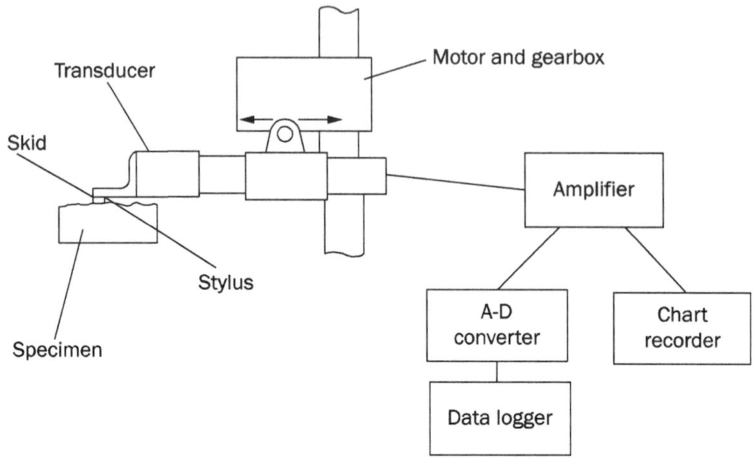

under examination. The transducer produces an electrical signal, proportional to displacement of the stylus, which is amplified and fed to a chart recorder that provides a magnified view of the original profile. But this graphical representation differs from the actual surface profile because of difference in magnifications employed in vertical and horizontal directions. Surface slopes appear very steep on profilometric record though they are rarely steeper than 10° in actual cases. The shape of the stylus also plays a vital role in incorporating error in measurement. The finite tip radius (typically 1 to 2.5 microns for a diamond stylus) and the included angle (of about 60° for pyramidal or conical shape) result in preventing the stylus from penetrating fully into deep and narrow valleys of the surface and thus some smoothing of the profile is done. Some error is also introduced by the stylus in terms of distortion or damage of a very delicate surface because of the load applied on it. In such cases a non-contacting optical profilometer having optical heads replacing stylus may be used. Reflection of infrared radiation from the surface is recorded by arrays of photodiodes and analysis of these in a microprocessor results in the determination of the surface topography. Vertical resolution of the order of 0.1nm is achievable though maximum height of measurement is limited to a few microns. This method is clearly advantageous in the case of very fine surface features.

5.2.2 Fractal characterisation

Fractal geometry

Euclidean geometry describes ordered objects such as points, curves, surfaces and cubes using integer dimensions of 0, 1, 2, and 3, respectively. A measure of the object such as the length of a line, the area of a surface and the volume of a cube are associated with each dimension. These measures are invariant with respect to the unit of measurement. It means that the length of a line remains independent of whether a centimetre or a micrometre scale is used. However, a multitude of objects found in nature appear disordered and irregular for which the measures of length, area and volume are scale-dependent. This suggests that the dimensions of such objects cannot be integers. A generalised concept of a dimension and the origins of fractal geometry are now discussed.

Mandelbrot (1967) founded fractal geometry when he showed that for decreasing the unit of measurement, the length of a natural coastline does not converge but, instead, increases monotonically. On plotting the length L as a function of the unit of measurement ϵ on a log-log plot, he found a simple relation of the form $L \sim \epsilon^{1-D}$. Mandelbrot finally made an interesting conclusion that the real number D associated with every coastline is the dimension of the coastline. This study marked the origins of fractal geometry, which has now found numerous applications in characterising and describing disordered phenomena in science and engineering.

Fractal dimension

To measure the length of a line, let us break the line into small units of length ϵ and then add the number of units in the form

$$L = \sum \epsilon^1 \tag{5.1}$$

Similarly to measure the area of surface, let us break up the surface into small squares of size $\epsilon \times \epsilon$ and then add the number of units as

$$A = \sum \epsilon^2 \tag{5.2}$$

Here in Equations 5.1 and 5.2 the exponents 1 and 2 correspond to the dimensions of the objects. These measures of length and area have a unique property that they are independent of the unit of measurement ϵ and in the limit $\epsilon \rightarrow 0$ these measures remain finite and non-zero. This concept of Euclidean dimension thus can be generalised in the form

$$M = \sum \epsilon^D \tag{5.3}$$

Here M is the measure and D is a real number. If the exponent D makes the measure M independent of the unit of measurement ϵ in the limit of $\epsilon \rightarrow 0$, then D is the dimension of an object.

Contrary to common understanding of dimension, this generalisation allows the dimension of an object to take non-integer values. If, in this argument, it is assumed that an object is broken into N equal parts then Equation 5.3 can be written as $M = N\epsilon^{D}$. Since the measure is invariant with the unit of measurement, one can write $N \sim \epsilon^{-D}$. Now if the length of an object is evaluated, then the length would vary as $L = N\epsilon^{1} \sim \epsilon^{1-D}$, as was observed for the lengths of the coastlines. It can be easily seen that the length will be independent of ϵ only when $D = 1$.

Self-similarity

The generalised concepts of measure and dimension are fundamental to the issue of self-similarity. Let us consider a one-dimensional line of unit length and break it up into N equal segments. Each segment of the line, of size 1/m, is similar to the whole line and needs a magnification of m to be an exact replica of the whole line. Since the length of the line remains independent of 1/m, it follows that the number of units is $N \sim m$. Now let us consider a square, which has a side of unit length. Each small square of side 1/m is similar to the whole square and needs a magnification of m to be an exact replica of the whole square. However, the number of small squares in the whole is $N \sim m^2$. In general, for an object of dimension D, it follows that

$$N \sim m^{D} \tag{5.4}$$

Thus the dimension of the object can be written as

$$D = \log N / \log m \tag{5.5}$$

This definition of dimension, which is based on the self-similarity of an object, is called the similarity dimension. To perceive what an object of a non-integer dimension looks like, one can follow the recursive construction in Figure 5.4, which yields the Koch curve of dimension 1.26.

In this construction the first step is to break a straight line into three parts and replace the middle portion by two

Figure 5.4 Formation of Koch curve

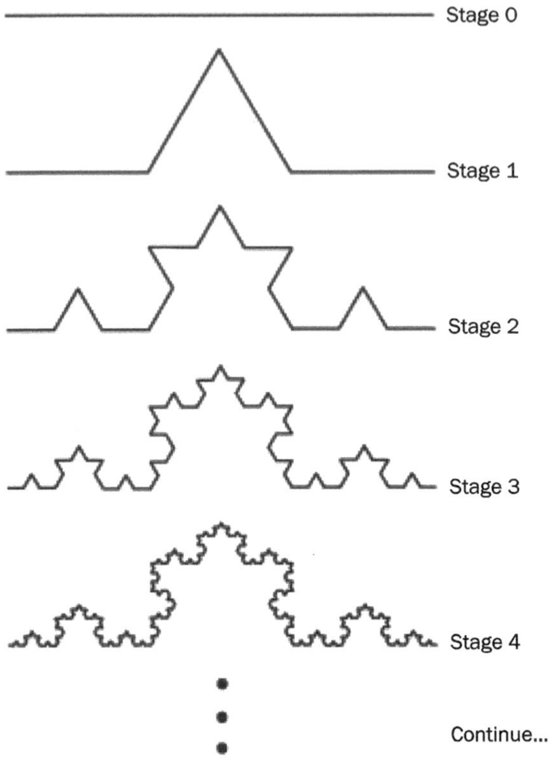

segments of equal lengths. In the subsequent stages each straight segment is broken into three parts and the middle portion of each segment is replaced by two parts. If this recursion is continued infinite times then the Koch curve is obtained. This curve has some unique mathematical properties. Firstly, the curve is continuous but it is not differentiable anywhere. The non-differentiability arises because of the fact that if the curve is repeatedly magnified, more and more details of the curve keep appearing. This means that tangent cannot be drawn at any point and therefore the curve cannot be differentiated. Secondly, the curve is exactly self-similar. This is because if a small portion of the curve is appropriately magnified, it will be an exact replica of the whole Koch curve. Thirdly, the dimension of the curve remains constant at all scales, although the curve contains roughness at a large number of scales. This scale-invariance of the dimension is an important property, which is utilised to characterise rough surfaces. The coastline of an island is an example of a self-similar object found in nature. Although these objects are not exactly self-similar, they are statistically self-similar. Statistical self-similarity means that the probability distribution of a small part of an object will be congruent with the probability distribution of the whole object if the small part is magnified appropriately. However, not all fractal objects are self-similar. This leads to the more general concept of self-affinity.

Self-affinity

The definition of self-similarity is based on the property of equal magnification in all directions. However, there are many objects in nature which have unequal scaling in different directions. Thus these are not self-similar but self-affine. The dimension of self-affine fractals cannot be obtained from Equation 5.5, which is based on the self-similarity of an object. Mandelbrot showed that the lengths of self-affine fractal curves do not follow the relation $L \sim \epsilon^{1-D}$ for all values of ϵ and therefore the dimension of self-affine curves cannot be obtained by measuring their lengths. Instead, the dimension of self-affine functions can be obtained from their power spectra.

Fractal description of roughness

The deviation of a surface from its mean plane is assumed to be a random process, which is characterised by the statistical parameters such as the variance of the height, the slope and the curvature. But, it has been observed that surface topography is a non-stationary random process. It means the variance of the height distribution is related to the sampling length and hence is not unique for a particular surface. Rough surfaces are also known to exhibit the feature of geometric self-similarity and self-affinity, by which similar appearances of the surface are seen under the various degrees of magnification as quantitatively shown in Figure 5.5. Since increasing amounts of detail in the roughness are observed at decreasing length scale, the concepts of slope and curvature, which inherently assume the smoothness of the surface, cannot be defined. So the variances of slope and curvature depend strongly on the resolution of the roughness-measuring instrument or some other form of filter and are therefore not unique. In contemporary literature such a large number of characterisation parameters occurs that the term 'parameters rash' is aptly used. The use of instrument-dependent parameters shows different values for the same surface. Thus, it is necessary to characterise rough surfaces by intrinsic parameters, which are independent of all scales of roughness. This suggests the use of fractal geometry in characterising the surface roughness. The fractal dimension is an intrinsic property and should be used for surface characterisation. It is invariant with length scales and is closely linked to the concept of geometric self-similarity.

Figure 5.5 Qualitative description of statistical self-affinity for a surface profile

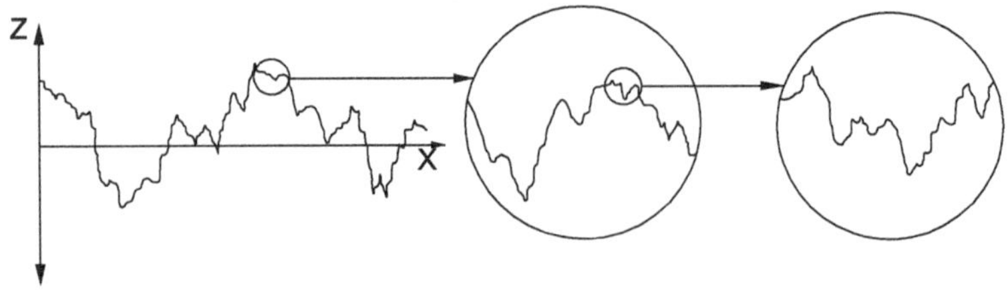

The self-similarity or self-affinity of rough surfaces implies that as the unit of measurement is continuously decreased, the surface area of the rough surface (a two-dimensional measure) tends to infinity and the volume (a three-dimensional measure) tends to zero. Here, self-similarity implies the property of equal magnification in all directions while self-affinity refers to unequal scaling in different directions. Thus, the Hausdorff or fractal dimension, $D + 1$, of rough surfaces is a fraction between 2 and 3. The profile of a rough surface $z(x)$, typically obtained from stylus measurements, is assumed to be continuous even at the smallest scales. This assumption breaks down at atomic scale. But for engineering surfaces the continuum is assumed to exist down to the limit of a zero-length scale. Since repeated magnifications reveal the finer levels of detail, the tangent at any point cannot be defined. Thus the surface profile is continuous everywhere but non-differentiable at all points. This mathematical property of continuity, non-differentiability and self-affinity (Berry and Lewis, 1980) is satisfied by the modified Weierstrass-Mandelbrot (W-M) fractal function, which is thus used to characterise and simulate such profiles. The W-M function has a fractal dimension D, between 1 and 2, and is given by

$$z(x) = G^{(D-1)} \sum_{n=n_1}^{\alpha} \frac{\cos 2\pi\gamma^{n} x}{\gamma^{(2-D)n}}, \quad 1<D<2,\ \gamma>1 \tag{5.6}$$

where G is a scaling constant. The parameter n_1 corresponds to the low cut-off frequency of the profile. Since surfaces are non-stationary random process the lowest cut-off frequency depends on the length L of the sample and is given by $\gamma^{n1} = 1/L$.

The W-M function has the interesting mathematical property that the series for $z(x)$ converges but that for dz/dx diverges. It implies that it is non-differentiable at all points.

The power spectrum of this W-M function can be expressed by a continuous function as

$$S(\omega) = \frac{G^{2(D-1)}}{2\ln\gamma}\frac{1}{\omega^{5-2D}} \tag{5.7}$$

When this equation is compared with the power spectrum of a surface, the dimension D is related to the slope of the spectrum on a log-log plot against ω. The constant G is the roughness parameter of a surface, which is invariant with respect to all frequencies of roughness and determines the position of spectrum along the power axis. In this characterisation method both G and D are independent of the roughness scales of the surface and hence intrinsic properties. The constants of the W-M function, G, D, and n_1 form a complete and fundamental set of scale-independent parameters to characterise a rough surface. The physical significance of D is the extent of space occupied by the rough surface, i.e. larger D values correspond to denser profile or smoother topography (Yan and Komvopoulos, 1998; Sahoo and Ghosh, 2007).

Fractal dimension calculation

Fractal calculation mainly includes the calculation of profile fractal dimension ($1<D<2$) and the calculation of surface fractal dimension ($2<D<3$) in tribological fractal research. Fractal calculation is generally involved with computer-assisted image analysis of topography images in 2-D or 3-D of a surface obtained in analogue or digital signals using profilometer or microscopy, etc. An effective method to convert these signals into the required data for calculating fractal dimensions must therefore be sought. Profile instruments can be used to obtain data in 2-D, which are then directly used to calculate fractal dimension.

The methods for calculating profile fractal dimension mainly include the yardstick, the box counting, the variation, the structure function and the power spectrum method (Sahoo, 2005).

The yardstick method employs the technique of 'walking' around a profile curve in a step length, *r*. A point on the profile curve is chosen as a starting point of divider, while another point at a distance *r* from the starting point is taken as its end point. Repetitively, find the point-pair of dividers in the same way until the profile curve is entirely measured. Then, the summing up of the step lengths enables the curve length to be determined. The repetition of this calculation process at various step lengths allows all the curve length to be evaluated. Further, plotting of the curve lengths versus the step lengths on a log-log scale gives the slope m of a fitting line to be related to the fractal dimension D as $D = 1 - m$. It is possible that this method has abandoned some pivotal points, resulting in calculation error.

The principle of the box counting method mainly involves an iteration operation to an initial square, whose area is supposed to be 1 and which covers the entire graph. The initial square is divided into four sub-squares and so on. After the n times operations, the number of sub-squares, which contain the discrete points of the profile graph are counted and the length L of the profile is approximately obtained. Then the fractal dimension is calculated as $D=1+\log L/(n.\log 2)$.

The variation method has the advantage of being proven theoretically for all profiles (self-affine or not), and of giving quickly an estimation of the dimension of mathematical profiles. A well-known technique used to analyse surfaces consists in performing 'slices' through the surfaces, which allows one to transform a 3-D problem to 2-D. In other words, a surface is replaced by profiles, taken at different places, and the fractal dimension estimated over profiles is then related to the 3-D fractal dimension by the classical result: dimension of surface = 1 + dimension of profiles. Such a technique obviously decreases the problem size. Accurate results are hard to obtain for the surface dimension and the variation method gives the best approximations. The variation method algorithm is based on the local oscillation of the profile function *Z*.

The power spectrum method involves the evaluation of the power of the profile function. The modified Weierstrass-Mandelbrot (W-M) function for a rough surface is described by Equation 5.6. The multi-scale nature of $z(x)$ can be characterised by its power spectrum, which gives the amplitude of the roughness at all length scales. The parameters G and D can be found from the power spectrum of the W-M function given by Equation 5.7. Usually, the power law behaviour would result in a straight line if $S(\omega)$ is plotted as a function of ω on a log-log graph. Using fast Fourier transform (FFT), the power spectrum of profile can be calculated and then be plotted versus the frequency on a log-log scale. Thereafter, the fractal dimension, D, can be related to the slope m of a fitting line on a log-log plot as: $D = (5 + m)/2$.

The structure function method considers all points on the surface profile curve as a time sequence $z(x)$ with fractal character. The structure function $s(\tau)$ of sampling data on the profile curve can be described as $s(\tau) = [z(x + \tau) - z(x)]^2 = c\tau^{4-2D}$ where $[z(x+\tau)-z(x)]^2$ expresses the arithmetic average value of difference square, and τ is the random choice value of data interval. Different τ and the corresponding $s(\tau)$ can be plotted versus the τ on a log-log scale. Then, the fractal dimension D can be related to the slope m of a fitting line on log-log plot as: $D = (4 - m)/2$.

Fractal dimension measurement in the present study

Roughness profile measurement is done using a stylus-type profilometer, *Talysurf* (Taylor Hobson, UK). The profilometer is set to a cut-off length of 0.8mm, Gaussian filter, traverse speed 1mm/sec and 4mm traverse length. Roughness measurements, in the transverse direction, on the work-pieces are repeated four times and the average of four measurements of surface roughness parameter values is recorded. The measured profile is digitised and processed through the dedicated advanced surface finish analysis software *Talyprofile*. Then fractal dimension is evaluated following the structure function method.

5.2.3 Design of experiments

The design of experiments technique is a very powerful tool, which permits carrying out the modelling and analysis of the influence of process variables on the response variables. The response variable is an unknown function of the process variables, which are known as design factors. The purpose of running experiments is to characterise unknown relations and dependencies that exist in the observed design or process, i.e. to find out the influential design variables and the response to variations in the design variable values. A scientific approach to planning the experiment must be employed if an experiment is to be performed most efficiently. The statistical design of experiments refers to the process of planning the experiment so that appropriate data that can be analysed by statistical methods will be collected, resulting in valid and objective conclusions in a meaningful way. When the problem involves data that are subject to experimental errors, statistical methodology is the only objective approach to analysis.

Sometimes, experiments are repeated with a particular set of levels for all the factors to check the statistical validation and repeatability by the replicate data. This is called replication. To get rid of any bias, allocation of experimental material and the order of experimental runs are randomly selected. This is called randomisation. To arrange the experimental material into groups, or blocks, that should be more homogeneous than the entire set of material is called blocking. So, when experiments are carried out these things should be remembered. There are several methodologies for design of experiments. Some DOE methods are presented below.

Full factorial design

Full factorial design creates experimental points using all the possible combinations of the levels of the factors in each complete trial or replication of the experiments. The experimental design points in a full factorial design are the vertices of a hyper cube in the n-dimensional design space defined by the minimum and the maximum values of each of the factors. These experimental points are also called factorial points. For three factors having four levels of each factor, considering full factorial design, total 4^3 (64) numbers of experiments have to be carried out. If there are n replicates of complete experiments, then there will be n times of the single replication experiments to be conducted. In the experimentation, it must have at least two replicates to determine a sum of squares due to error if all possible interactions are included in the model.

Central composite design

A Box-Wilson central composite design, commonly called central composite design (CCD), is frequently used for building a second-order polynomial for the response variables in response surface methodology without using a complete full factorial design of experiments. To establish the coefficients of a polynomial with quadratic terms, the experimental design must have at least three levels of each factor. In CCD, there are three different points, namely factorial points, central points and axial points. Factorial points are vertices of the n-dimensional cube which are coming from the full or fractional factorial design where the factor levels are coded to –1, +1. Central point is the point at the centre of the design space. Axial points are located on the axes of the coordinate system symmetrically with respect to the central point at a distance α from the design centre.

There are two main varieties of CCD, namely face-centred CCD and rotatable CCD. In face-centred CCD, a k factor three-level experimental design requires 2^k+2k+C experiments, where k is the number of factors, 2^k points are in the corners of the cube representing the experimental domain, $2k$ axial points are in the centre of each face of the cube $[(\pm\alpha, 0,...0), (0, \pm\alpha,...0),... (0, 0,...\pm\alpha)]$ and C points are the replicates in the centre of the cube that are necessary to estimate the variability of the experimental measurements, that is to say, the repeatability of the phenomenon which carries out the lack-of-fit or curvature test for the model. The centre points may vary from three to six. The example of three-level three-factor FCC design is shown in Figure 5.6. In this figure the deep black circles represent the fractional points at the corner of the cube while the white circles represent axial points in the centre of each face of the cube and the star mark represents the centre points. For the three-factor experiment, eight (2^3) factorial points, six axial points (2×3) and six centre runs, a total of 20 experimental runs can be considered. The value of α is chosen here as 1. The upper and lower limits of a factor are coded as +1 and –1 respectively using the following relations

Figure 5.6 Face-centred central composite design with three factors

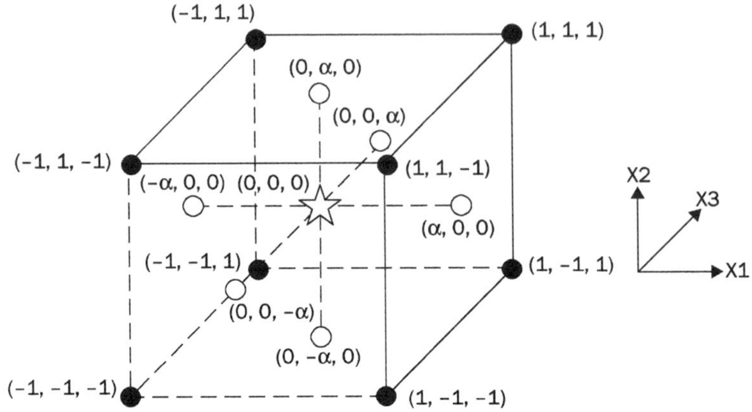

(Equation 5.8). Generally, the experimental runs are conducted in random order.

$$x_i = \frac{[2x - (x_{max} + x_{min})]}{(x_{max} - x_{min})} \tag{5.8}$$

The rotatable central composite design is the most widely used experimental design for modelling a second-order response surface. A design is called rotatable when the variance of the predicted response at any point depends only on the distance of the point from the centre point of design. The rotatable design provides the uniformity of prediction error and it is achieved by proper choice of α. In rotatable designs, all points at the same radial distance (r) from the centre point have the same magnitude of prediction error. For a given number of variables, the α required to achieve rotatability is computed as $\alpha = (n_f)^{1/4}$, where n_f is the number of points in the 2^k factorial design. A rotatable CCD consists of 2^k fractional factorial points, augmented by 2k axial points $[(\pm\alpha, 0,\ldots 0),(0, \pm\alpha,\ldots 0), (0, 0, \ldots \pm\alpha)]$ and n_c centre points (0, 0, 0, 0. . .,0). Here also, the centre points vary from three to six. With proper choice of n_c the CCD can be made orthogonal or it can be made uniform precision design. It means that the variance of response at origin is equal to the variance of response at a unit distance from the origin. Considering uniform precision, for three-factor experimentation, eight (2^3) factorial points, six axial points (2×3) and six centre runs, a total of 20 experimental runs may be considered and the value of α is $(8)^{1/4} = 1.682$.

5.2.4 Artificial neural network

A neural network is a massively parallel distributed processor that has a natural propensity for storing experimental knowledge and making it available for use (Haykin, 2007). Artificial neural networks are simplified models of biological nervous system which mimic the computational architecture of a human brain. In the human brain there are about 10 million interconnected neurons and these neurons are the core components of the nervous system. In the same way, an artificial neural network comprises processing elements called neurons. These neurons are connected via connection links that carry signals between processors and generally grouped into an input layer, a variable number of hidden layers and an output layer. The input layer receives the information from an external source. The hidden layer will receive information from the input layer and create internal representation. Finally, output will be available via the output layer. The network will acquire knowledge through a learning process. The learning is done by examples. The connection links possess some weights and these synaptic weights are used to store the knowledge in the network. The weights are adjusted during the learning stage to get the desired result for a given set of inputs. The greater the weight between neurons, the stronger the influence of the connection.

There are several learning methods developed for ANN. The back propagation learning method is frequently used by many researchers successfully in manufacturing sciences. This learning method is simple, most representative and commonly used. The basic structure of a back propagation neural network consists of an input layer, hidden layer(s) and an output layer. There may be one or more hidden layers consisting of variable neurons or nodes. The basic neural architecture is presented in Figure 5.7.

The input layer is used to feed the data in the network. The inputs are subsequently modified based on the interconnection weights between the layers. The net input to each neuron from the preceding layer will be as

$$net_j = \sum_{i=1}^{N} w_{ij} x_i + b_j \qquad (5.9)$$

where net_j is the net input, N is the number of neurons of the inputs to the jth neuron in the hidden layer, w_{ij} is the interconnected weight from i^{th} neuron in the forward layer

Figure 5.7 Architecture of the neural network

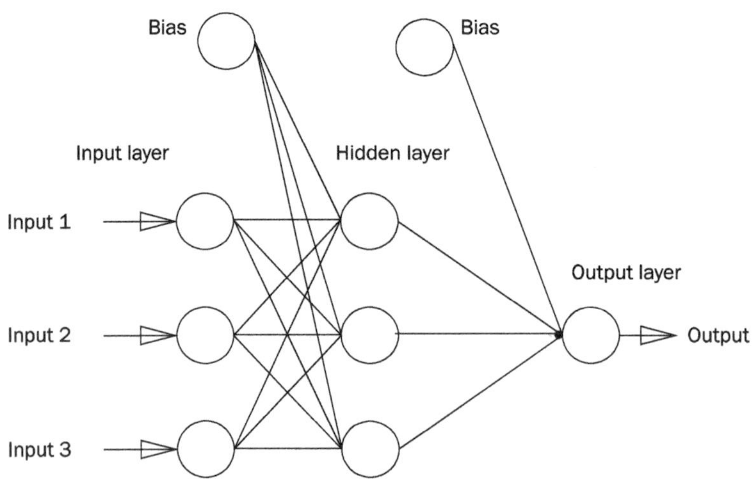

to j^{th} neuron in the hidden layer, x_i is the input from the ith neuron in the hidden layer and b_j is the bias value to the corresponding neuron. The output signal is obtained by applying activations to the net input. To construct the models, generally, the hyperbolic tangent sigmoid function in the hidden layer(s) and linear activation function in the output layer is considered.

The performance of a particular problem in neural network depends on various factors such as number of training data sets, performance goal, complexity of the data set, network architecture, which includes the number of neurons in the hidden layers and the number of layers, etc.

Training of the network

There are many training algorithms. But it is very difficult to predict which training algorithm will provide the best performance. Standard back-propagation is a gradient descent algorithm in which the network weights and biases update along the negative of the gradient of the performance function. The problem with standard gradient descent algorithm is that it converges very slowly. With standard steepest descent, the learning rate is held constant throughout training. But the performance of the algorithm is very sensitive to the proper setting of the learning rate. The performance of the algorithm can be improved using adaptive learning rate. The learning rate is made responsive to the complexity of the local error surface. Newton's method is an alternative to the conjugate gradient methods for fast optimisation. The basic step of Newton's method is to compute the Hessian matrix (second derivatives) of the performance index at the current values of the weights and biases and the Hessian matrix of second derivatives of the function to be minimised. Newton's method often converges faster than conjugate gradient methods. But it is complex and expensive to compute the Hessian matrix for feed-forward neural networks. Levenberg-Marquardt can approach second-order training speed without having to compute the Hessian matrix (second derivatives) (Demuth and Beale, 1998).

Initialisation of weights

Initialisation of weights in the network has great impact on the performance of the network. The Nguyen-Widrow weight initialisation algorithm is applied which generates initial weight and bias values for a layer so that the active regions of the layer's neurons will be distributed roughly evenly over the input space (Demuth and Beale, 1998).

Generalisation of the models

The main target of a properly trained neural network is to give reasonable answers when presented with inputs that they have never seen. That means the network should be generalised well to take new inputs. There are different techniques to improve generalisation, namely early stopping, Bayesian regularisation, etc. In the 'early stopping' technique, the complete set of data is divided into three subsets, namely training set, testing set and validation set. The training set is used to train the network. The validation set determines when the training should stop by monitoring the error. The testing set does not participate in the training of the network but is used to test the generalisation of the trained network. Bayesian regularisation (MacKay, 1992; Foresee and Hagan, 1997) generally provides better generalisation performance than early stopping, when training function approximation networks. Bayesian regularisation generally does not require a separate validation set; it uses all data points of training sets for validation purpose also (Demuth and Beale, 1998). The Bayesian regularisation algorithm updates the weight and bias values according to Levenberg-Marquardt optimisation. It minimises a combination of squared errors and weights, and then determines the correct combination so as to produce a network that generalises well.

Selection criterion of the best network

The determination of hidden layer(s) and numbers of neurons in the hidden layer(s) is a considerable task. But, generally, one hidden layer network is selected first. If it does not perform well then two hidden layers are used. For selecting numbers of neurons in the hidden layer, many researchers propose different algorithms/criteria. One possible way of selection of neurons in the hidden layers is by increasing the number of neurons to find out the best neural network model. There are several criteria for selecting the best network. But, generally, minimum mean squared error (MSE) is used as the selection parameter. Mean squared error is measured as

$$MSE = \frac{1}{N} \sum_{p=1}^{N} \sum_{k=1}^{n} (d_{k,p} - o_{k,p})^2 \qquad (5.10)$$

where N is the number of pattern, n is the number of node in the output layer, $d_{k,p}$ is the desired/experimental output for k^{th} node of the pth pattern and $o_{k,p}$ is the calculated/predicted output of kth node of pth pattern.

Besides this parameter, correlation coefficient (R) and maximum absolute percentage error for the network are also selection parameters. Correlation coefficient (R) is an indication of agreement between experimental and predicted values. It ranges from −1.00 to +1.00. Correlation coefficient of −1.00 represents a perfect negative correlation whereas correlation coefficient of +1.00 represents a perfect positive correlation. Absolute percentage error for the training and testing data set is measured as

$$Absolute\_Percentage\_error = \left| \frac{y_{i,experimental} - y_{i,predicted}}{y_{i,experimental}} \right| \times 100 \quad (5.11)$$

where $y_{i,experimental}$ is the experimental value and $y_{i,predicted}$ is the network predicted value of the response for i^{th} observation.

5.3 CNC end milling

CNC milling is a popular machining process in the modern industry because of its ability to remove materials with a multi-point cutting tool at a faster rate with a reasonably good surface quality. In order to get specified surface roughness, selection of controlling parameters is necessary. In CNC milling there are several parameters which control the surface quality. The analysis of surface roughness on CNC end milling process is a big challenge for research development. Several factors involved in the machining process have to be optimised to obtain a desired surface quality. In this study, three machining parameters are considered, namely spindle speed, feed rate and depth of cut. To develop the surface roughness prediction model in terms of fractal dimension, neural network is used.

5.3.1 *Experimental details*

Design of experiments

A full factorial design is used with five levels of each of the three design factors, namely depth of cut (*d*, mm), spindle speed (N, rpm) and feed rate (f, mm/min). Thus the design chosen was five level-three factor (5^3) full factorial design consisting of 125 sets of coded combinations for each work-piece material. Three cutting parameters are selected as design factors while other parameters have been assumed to be constant over the experimental domain. The upper and lower limits of a factor were coded as +1 and –1 respectively using Equation 5.8. The process variables/design factors with their values on different levels are listed in Table 5.1.

Table 5.1 Variable levels used in CNC milling

	Parameters		
Levels	*d*	*N*	*f*
–1	0.150	2,500	300
–0.5	0.175	2,750	350
0	0.200	3,000	400
0.5	0.225	3,250	450
1	0.250	3,500	500

Machine used

The machine used for the milling tests is a 'DYNA V4.5' CNC end milling machine having the control system SINUMERIK 802 *D* with a vertical milling head. The specification of CNC end milling machine is shown in Table 5.2. For generating the milled surfaces, CNC part programmes for tool paths were created with specific commands. The compressed coolant servo-cut was used as cutting environment.

Table 5.2 **Specification of CNC end milling machine**

Table size	450x250mm
Table load capacity	200kg
X travel	250mm
Y travel	175mm
Z travel	175mm
Spindle nose to table	300mm
Spindle centre to column	280mm
Taper of spindle nose	BT 30
Spindle speed	9,000rpm
Rapid on X and Y axis	15m/min
Rapid on Z axis	10m/min
Spindle motor	3.7kW
X axis motor	3Nm
Y axis motor	3Nm
Z axis motor	6Nm
Control system	802 *D* SINUMERIK
Power requirement	7.5kW/10HP
Lubricating oil	Tellus 33 or EN KLO 68

Cutting tools used

Coated carbide tools are known to perform better than uncoated carbide tools. Thus commercially available CVD coated carbide tools were used in this investigation. The tools used were flat end mill cutters produced by WIDIA (EM-TiAlN). The tools were coated with TiAlN coating. For each material a new cutter of same specification was used. The details of the end milling cutters are given below:

- cutter diameter = 8mm
- overall length = 108mm
- fluted length = 38mm

- helix angle = 30°
- hardness = 1,570HV
- density = 14.5g/cc
- transverse rupture strength = 3,800N/mm^2

Work-piece material

The present study was carried out with AISI 1040 steel. The chemical composition and mechanical properties of the work-piece material are shown in Table 5.3. All the specimens were in the form of 100mm x 75mm x 25mm blocks.

Table 5.3 **Composition and mechanical properties of work-piece material**

Work material	Chemical composition (%Wt)	Mechanical property
Mild Steel (AISI 1040)	0.42%C, 0.48%Mn, 0.17%Si, 0.02%P, 0.018%S, 0.1%Cu, 0.09%Ni, 0.07%Cr and balance Fe	Hardness-201 BHN, Density-7.85 g/cc, Tensile strength-620 MPa

5.3.2 Results and discussion

CNC milling operations are carried out on mild steel work-pieces to get machined surfaces for different combinations of spindle speed, feed rate and depth of cut. The generated surfaces are measured using the Talysurf instrument and further processed to get fractal dimension (D). The experimental results are used to develop the prediction model using artificial neural networks to model fractal dimension. For ANN, full factorial design of experiments is considered and the experimental results are presented in Table 5.4 (Sahoo et al., 2008).

According to the full factorial design of experiments, a total of 125 experiments are carried out. These datasets are randomised and 60% of data are chosen for training, 20% for

Table 5.4 **Experimental results for CNC milling considering full factorial design**

Sl No.	Depth of cut (*d*)	Spindle speed (*N*)	Feed rate (*f*)	*D*
1	–1	–1	–1	1.31
2	–1	–1	–0.5	1.33
3	–1	–1	0	1.29
4	–1	–1	0.5	1.30
5	–1	–1	1	1.32
6	–1	–0.5	–1	1.29
7	–1	–0.5	–0.5	1.33
8	–1	–0.5	0	1.37
9	–1	–0.5	0.5	1.37
10	–1	–0.5	1	1.34
11	–1	0	–1	1.32
12	–1	0	–0.5	1.35
13	–1	0	0	1.38
14	–1	0	0.5	1.34
15	–1	0	1	1.39
16	–1	0.5	–1	1.38
17	–1	0.5	–0.5	1.36
18	–1	0.5	0	1.36
19	–1	0.5	0.5	1.40
20	–1	0.5	1	1.34
21	–1	1	–1	1.40
22	–1	1	–0.5	1.38
23	–1	1	0	1.41
24	–1	1	0.5	1.36
25	–1	1	1	1.37
26	–0.5	–1	–1	1.41
27	–0.5	–1	–0.5	1.39
28	–0.5	–1	0	1.35
29	–0.5	–1	0.5	1.39
30	–0.5	–1	1	1.38

31	–0.5	–0.5	–1	1.31
32	–0.5	–0.5	–0.5	1.37
33	–0.5	–0.5	0	1.40
34	–0.5	–0.5	0.5	1.41
35	–0.5	–0.5	1	1.40
36	–0.5	0	–1	1.38
37	–0.5	0	–0.5	1.32
38	–0.5	0	0	1.37
39	–0.5	0	0.5	1.39
40	–0.5	0	1	1.39
41	–0.5	0.5	–1	1.41
42	–0.5	0.5	–0.5	1.40
43	–0.5	0.5	0	1.36
44	–0.5	0.5	0.5	1.41
45	–0.5	0.5	1	1.36
46	–0.5	1	–1	1.38
47	–0.5	1	–0.5	1.38
48	–0.5	1	0	1.32
49	–0.5	1	0.5	1.39
50	–0.5	1	1	1.41
51	0	–1	–1	1.38
52	0	–1	–0.5	1.42
53	0	–1	0	1.43
54	0	–1	0.5	1.43
55	0	–1	1	1.41
56	0	–0.5	–1	1.38
57	0	–0.5	–0.5	1.41
58	0	–0.5	0	1.38
59	0	–0.5	0.5	1.38
60	0	–0.5	1	1.40
61	0	0	–1	1.38
62	0	0	–0.5	1.37
63	0	0	0	1.34

(*Continued*)

Table 5.4 *Continued*

Sl No.	Depth of cut (*d*)	Spindle speed (*N*)	Feed rate (*f*)	*D*
64	0	0	0.5	1.41
65	0	0	1	1.38
66	0	0.5	–1	1.40
67	0	0.5	–0.5	1.39
68	0	0.5	0	1.36
69	0	0.5	0.5	1.40
70	0	0.5	1	1.38
71	0	1	–1	1.43
72	0	1	–0.5	1.41
73	0	1	0	1.40
74	0	1	0.5	1.39
75	0	1	1	1.43
76	0.5	–1	–1	1.40
77	0.5	–1	–0.5	1.39
78	0.5	–1	0	1.38
79	0.5	–1	0.5	1.43
80	0.5	–1	1	1.38
81	0.5	–0.5	–1	1.39
82	0.5	–0.5	–0.5	1.35
83	0.5	–0.5	0	1.37
84	0.5	–0.5	0.5	1.40
85	0.5	–0.5	1	1.41
86	0.5	0	–1	1.35
87	0.5	0	–0.5	1.32
88	0.5	0	0	1.37
89	0.5	0	0.5	1.39
90	0.5	0	1	1.41
91	0.5	0.5	–1	1.36
92	0.5	0.5	–0.5	1.38
93	0.5	0.5	0	1.37

94	0.5	0.5	0.5	1.39
95	0.5	0.5	1	1.38
96	0.5	1	–1	1.44
97	0.5	1	–0.5	1.43
98	0.5	1	0	1.44
99	0.5	1	0.5	1.43
100	0.5	1	1	1.42
101	1	–1	–1	1.30
102	1	–1	–0.5	1.42
103	1	–1	0	1.38
104	1	–1	0.5	1.38
105	1	–1	1	1.39
106	1	–0.5	–1	1.35
107	1	–0.5	–0.5	1.38
108	1	–0.5	0	1.33
109	1	–0.5	0.5	1.36
110	1	–0.5	1	1.40
111	1	0	–1	1.39
112	1	0	–0.5	1.37
113	1	0	0	1.35
114	1	0	0.5	1.39
115	1	0	1	1.41
116	1	0.5	–1	1.40
117	1	0.5	–0.5	1.38
118	1	0.5	0	1.38
119	1	0.5	0.5	1.36
120	1	0.5	1	1.39
121	1	1	–1	1.41
122	1	1	–0.5	1.41
123	1	1	0	1.40
124	1	1	0.5	1.40
125	1	1	1	1.36

testing and 20% for validation purpose. Three cutting parameters, namely spindle speed, feed rate and depth of cut, are used as the input neurons and corresponding fractal dimension as the output neuron. The data points are normalised so that the datasets range in –1 to +1. To construct the models, the hyperbolic tangent sigmoid function in the hidden layer and linear activation function in the output layer are considered. To train the network, Levenberg-Marquardt (L-M) training algorithm is applied. Initial learning rate for the network is 0.001 with learning rate decrease factor as 0.10 and learning rate increase factor as 10. The maximum learning rate is 10^{10}. Initialisation of weights in the network has been done using Nguyen-Widrow weight initialisation algorithm. The number of neurons in the hidden layer is decided by trial with increasing number of neurons. The neurons in the hidden layers are varied from 5 to 20 to find out the best neural network model. Minimum mean squared error (MSE) is used as the selection parameter. Besides this parameter, for the network, correlation coefficient (R) and maximum absolute percentage error for the training data set and testing data set are calculated. MATLAB 6.5 is used to develop the neural network, train and test the network.

The performances of different developed networks are presented in Table 5.5. It is seen from the table that increasing the number of neurons in the hidden layer does not always ensure that the mean squared error will decrease but it is clear that by increasing the number of neurons, the architecture gets complicated. The architecture 3-17-1 gives the minimum mean squared error and is selected as the best performing architecture. This network also provides the highest correlation coefficient and minimum mean absolute percentage error.

From Table 5.5, it is seen that the maximum absolute percentage error is 2.01% for training data set which implies that the ANN predicted and experimental fractal dimensions

Table 5.5 Performance comparison of different models in CNC milling

Model	MSE	Correlation coefficient (R)	Max. training error (%)	Max. testing error (%)
3-5-1	0.00055765	0.719	5.17	3.31
3-6-1	0.00038646	0.734	3.75	4.25
3-7-1	0.00042188	0.809	4.20	3.34
3-8-1	0.00037304	0.811	3.81	3.53
3-9-1	0.00029953	0.834	3.50	3.78
3-10-1	0.00031412	0.833	3.36	3.89
3-11-1	0.00029204	0.849	4.17	3.47
3-12-1	0.00027958	0.842	4.65	3.53
3-13-1	0.00027715	0.853	2.70	3.67
3-14-1	0.00025740	0.868	5.11	3.03
3-15-1	0.00026038	0.850	2.83	3.70
3-16-1	0.00020333	0.898	3.21	3.35
3-17-1	**0.00016643**	**0.909**	**2.01**	**2.66**
3-18-1	0.00025489	0.850	5.15	3.65
3-19-1	0.00026657	0.868	2.81	3.62
3-20-1	0.00019169	0.896	3.44	5.18

are close to each other. The performance of the neural network is tested by carrying out regression analysis and it is seen that correlation coefficient (R) is 0.909 (Figure 5.8), which is the good indication of correlation of predicted and experimental fractal dimensions. The network is also tested with 25 new testing sets. Table 5.6 presents the comparative study of the experimental fractal dimension (*D*) and ANN predicted fractal dimension (*D*) for testing pattern. Maximum absolute error and mean absolute error are 2.66 and 1.15 respectively for testing pattern (Table 5.6). It implies that the experimental value and ANN predicted values lie very close to each other. Regression analysis is conducted for testing pattern also and it is observed from the regression analysis that correlation coefficient (R) is 0.84 (Figure 5.9). From the

Figure 5.8 Performance of ANN model using regression analysis for training pattern (CNC milling)

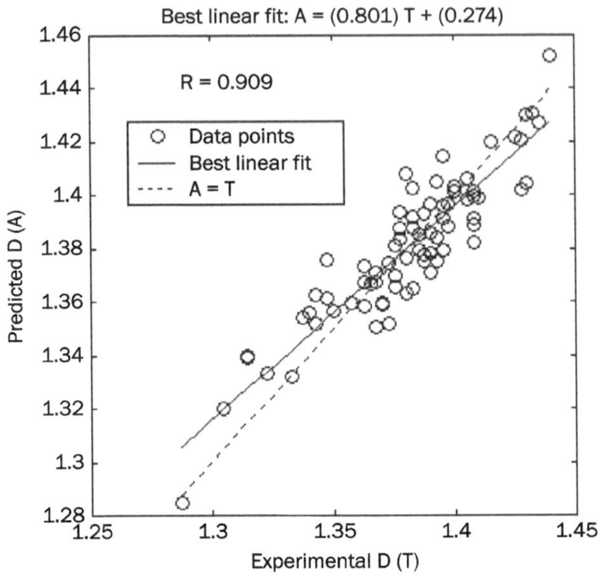

Table 5.6 Comparative study of experimental and ANN predicted fractal dimension, *D* for testing set (CNC milling)

Cutting parameters			Experimental D	Neural network predicted D	Absolute percentage error for NN
d	N	f			
–1	–1	0.5	1.30	1.2927	0.56
–1	–1	1	1.32	1.3078	0.92
–1	–0.5	–1	1.29	1.2698	1.57
–1	–0.5	–0.5	1.33	1.3195	0.79
–1	0.5	–0.5	1.36	1.3692	0.68
–1	0.5	0	1.36	1.3782	1.34
–1	1	1	1.37	1.3578	0.89
–0.5	–1	1	1.38	1.3899	0.72
–0.5	–0.5	0.5	1.41	1.3865	1.67
–0.5	0.5	1	1.36	1.3943	2.52

0	–1	1	1.41	1.4080	0.14
0	–0.5	1	1.40	1.4102	0.73
0	0.5	–1	1.40	1.4090	0.64
0	0.5	1	1.38	1.3723	0.56
0.5	–0.5	0.5	1.40	1.3926	0.53
0.5	0	–1	1.35	1.3770	2.00
0.5	0	0.5	1.39	1.3603	2.14
0.5	0.5	–1	1.36	1.3716	0.85
0.5	0.5	0	1.37	1.3605	0.69
1	–0.5	–0.5	1.38	1.3865	0.47
1	–0.5	0	1.33	1.3654	2.66
1	0.5	0	1.38	1.3741	0.43
1	0.5	1	1.39	1.3596	2.19
1	1	–0.5	1.41	1.3798	2.14
1	1	0.5	1.40	1.4122	0.87
Mean absolute error for ANN predicted D = 1.15%					

Figure 5.9 **Performance of ANN model using regression analysis for testing pattern (CNC milling)**

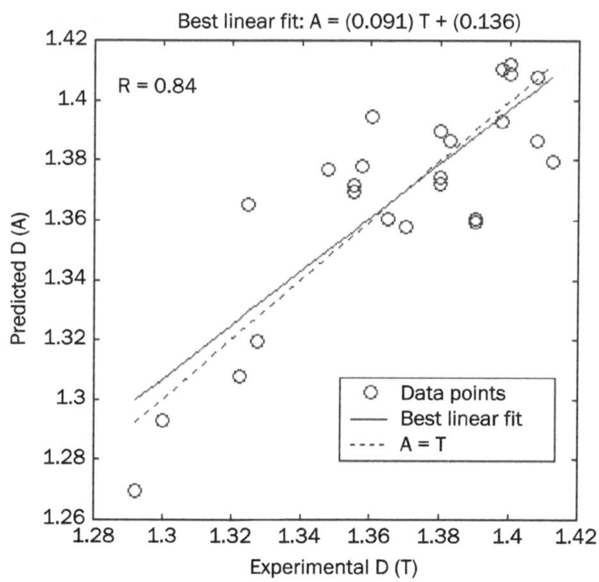

results, it is obvious that the network gives a good prediction of fractal dimension in milling of mild steel material.

5.4 CNC turning

Turning operation is an old and very common machining process in the industry. In recent times, the use of computer numerically controlled (CNC) machines has become popular to minimise the operator input and to get higher surface finish. In turning, there are several machining parameters which control the surface quality of the machined work-piece, which include cutting conditions, tool variables and work-piece variables. Cutting conditions include speed, feed and depth of cut, whereas tool variables include tool material, nose radius, rake angle, cutting edge geometry, tool vibration, tool overhang, tool point angle, etc. and work-piece variables include material hardness and other mechanical properties. It is very difficult to consider all the parameters that control the surface quality. But, the review of literature shows that the depth of cut, spindle speed and feed rate are the most widespread machining parameters taken by the researchers. In this study, CNC turning operations are carried out varying the machining parameters, namely depth of cut (mm), spindle speed (rpm) and feed rate (mm/rev). Machining surfaces are further analysed to find out the profile fractal dimension. These experimental results are further analysed using artificial neural network (ANN).

5.4.1 Experimental details

Design of experiments

For the experimentation, a rotatable central composite design is selected. The process variables with their values are listed in Table 5.7 and the experimental plan consists of experiment run order, standard order, coded values and actual values of process parameters as shown in Table 5.8.

Table 5.7 **Process parameter levels used in CNC turning**

Process variables	Unit	Levels				
		-1.682	**-1**	**0**	**1**	**1.682**
A						
Depth of cut(*d*)	mm	0.032	0.1	0.2	0.3	0.368
B						
Spindle speed(*N*)	rpm	528	800	1,200	1,600	1,872
C						
Feed rate(*f*)	mm/rev	0.0224	0.07	0.14	0.21	0.2576

Table 5.8 **Design matrix of the rotatable CCD design in CNC turning**

Std. order	Run order	Coded values			Actual values		
		d	*N*	*f*	*d*	*N*	*f*
1	20	–1	–1	–1	0.1	800	0.07
2	1	1	–1	–1	0.3	800	0.07
3	9	–1	1	–1	0.1	1,600	0.07
4	11	1	1	–1	0.3	1,600	0.07
5	7	–1	–1	1	0.1	800	0.21
6	8	1	–1	1	0.3	800	0.21
7	13	–1	1	1	0.1	1,600	0.21
8	3	1	1	1	0.3	1,600	0.21
9	10	–1.682	0	0	0.032	1,200	0.14
10	6	1.682	0	0	0.368	1,200	0.14
11	5	0	–1.682	0	0.2	528	0.14
12	14	0	1.682	0	0.2	1,872	0.14
13	12	0	0	–1.682	0.2	1,200	0.0224
15	2	0	0	0	0.2	1,200	0.14

(*Continued*)

Table 5.8 *Continued*

Std. order	Run order	Coded values			Actual values		
		d	*N*	*f*	*d*	*N*	*f*
16	4	0	0	0	0.2	1,200	0.14
17	17	0	0	0	0.2	1,200	0.14
18	16	0	0	0	0.2	1,200	0.14
19	18	0	0	0	0.2	1,200	0.14
20	15	0	0	0	0.2	1,200	0.14

Machine used

The machine used for the turning is a JOBBERXL CNC lathe having the control system FANUC Series Oi Mate-Tc and equipped with maximum spindle speed of 3,500rpm, feed rate 15–20m/rev and KVA rating –16 KVA. For generating the turned surfaces, CNC part programmes for tool paths were created with specific commands.

Cutting tool used

Coated carbide tools are known to perform better than uncoated carbide tools. Thus commercially available CVD coated carbide tools were used in this investigation. The tool holder is used as the PTGNR-25-25 M16 050, WIDIA and insert used as the TNMG 160404 –FL, WIDIA. The tool is coated with titanium nitride coating having hardness, density and transverse rupture strength as 1,570HV, 14.5g/cc and 3,800N/mm^2. The compressed coolant WS 50-50 with a ratio of 1:20 with water was used as cutting environment.

Work-piece material

The present study was carried out with mild steel (AISI 1040) work-pieces. All the specimens were in the form of bar with diameter 20mm and length 60mm.

5.4.2 Results and discussion

The generated machined surfaces are measured using the Talysurf instrument and further processed to get fractal dimension (*D*). The experimental results are presented in Table 5.9 (Barman and Sahoo, 2009) and used for further analyses using artificial neural networks to model fractal dimension.

Artificial neural models for predicting fractal dimension have been developed using multi-layer feed-forward back propagation algorithm. To construct the neural network,

Table 5.9 **Experimental results for CNC turning**

Std. order	D
1	1.370
2	1.300
3	1.362
4	1.410
5	1.267
6	1.282
7	1.390
8	1.417
9	1.320
10	1.370
11	1.300
12	1.420
13	1.360
14	1.290
15	1.397
16	1.400
17	1.415
18	1.415
19	1.402
20	1.412

work-piece speed, feed rate and depth of cut are used as input parameters of the network and corresponding fractal dimension as the output parameter. Only one hidden layer is used in the networks. The numbers of neurons are varied to select best network based on the minimum mean squared error (MSE). For training of the network, Levenberg-Marquardt algorithm is used. Since the CCD experimental scheme contains only 20 runs, all these observations are used for training the networks.

MSE values for different network architectures are presented in Table 5.10. From this table, it is seen that for mild steel material, increase in the number of neurons in the hidden layer beyond 4 does not improve the performance and thus network with 3-4-1 architecture is selected as the best network based on minimum MSE. For mild steel material, the maximum absolute percentage error is obtained as 0.69%. Thus, it is clear that the experimental values and ANN predicted values are very close. The comparative study of experimental fractal dimension and ANN model predicted fractal dimension is shown in Figure 5.10. From this figure also, it is clear that the predicted and experimental fractal dimension are very close to each other. The performance of the neural network is measured by carrying out the regression analysis. The correlation coefficient (R) for mild steel is 0.997 (Figure 5.11) and it is a good indication of the

Table 5.10 **MSE data for different architecture in CNC turning**

Serial No.	ANN architecture	MSE for mild steel
1	3-3-1	0.00003698
2	3-4-1	0.00001589
3	3-5-1	0.00001589
4	3-6-1	0.00001589

Figure 5.10 Comparative study of experimental *D* with ANN model predicted *D* in CNC turning

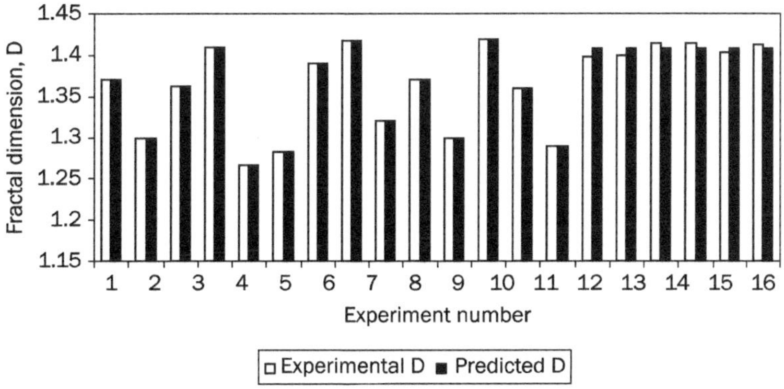

Figure 5.11 Performance of ANN model using regression analysis in CNC turning

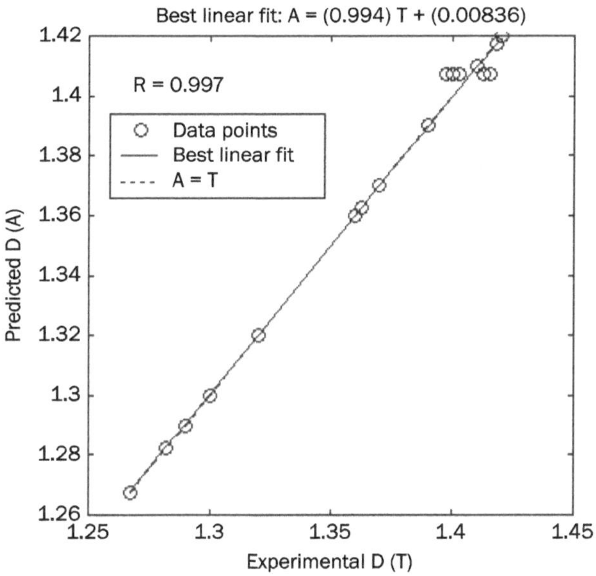

correlation between predicted and experimental outputs. Thus, it is seen that the ANN model predicts the fractal dimension in CNC turning of mild steel material with a good agreement with the experimental results.

5.5 Cylindrical grinding

Grinding is one of the common machining processes. In today's production, finishing of components is done by grinding due to the fact that it has the great potential to replace other machining processes and to achieve significant reduction in production time and cost. The acceptance of grinding as a finishing process is connected with a high form and size accuracy, high surface finish and surface integrity of the work-piece. In grinding there are several parameters which control the surface quality. In this study, only three machining parameters are considered, namely work-piece speed, longitudinal feed and radial infeed for machining mild steel (AISI 1040) work-pieces. The experimental results are analysed using artificial neural network. The experimental details and the results are discussed below.

5.5.1 Experimental details

Design of experiments

The process parameters chosen here are work-piece speed (N) in rpm, longitudinal feed (f) in mm/rev and radial infeed (d) in mm. The process variables/design factors with their values on different levels are listed in Table 5.11. The selection of the values of the variables is limited by the capacity of the machine used in the experimentation as well as the recommended specifications for different work-piece–tool material combinations. Four levels, having nearly equal spacing, within the operating range of the parameters are selected for each of the factors. By selecting four levels, the curvature or non-linearity effects can be studied. In the present investigation, full factorial design of experiment is considered for the experimentation and for

Table 5.11 **Process variables and their levels in grinding**

Parameters	Unit	Notation	1	2	3	4
Work-piece speed	rpm	*N*	56	80	112	160
Long feed	mm/rev	*f*	11.33	17.00	22.66	28.33
Radial infeed	mm	*d*	0.02	0.04	0.06	0.08

four level-three factors, a total of 64 experimental trials are carried out.

Machine used

The machine used for grinding is a HMT-made, K130U grinding machine equipped with maximum wheel speed of 1,910rpm. The wheel signature of the machine is A70K5V10 and wheel diameter is 270mm. The maximum grinding length is about 340mm. The compressed coolant WS 50-50 with a ratio of 1:20 was used as cutting environment. The details of the machine used in this study are shown in the Table 5.12.

Table 5.12 **Specification of the cylindrical grinding machine used**

Make	**HMT**	**Model**	**K130U**	**Machine No.**	**57169**
Maximum grinding length			340mm		
Maximum distance between centers			340mm		
Maximum travel of the table			310mm		
Maximum swivel of the table			200mm		
Grinding wheel					
Wheel speed			1,910rpm and 2,120rpm		
Wheel signature			A70K5V10		
Wheel diameter			270mm		

(Continued)

Table 5.12 *Continued*

Make	HMT	Model	K130U	Machine No	57169
Face width			40mm		
Bore diameter			50mm		
Work head					
Number of speed			8 (56-80-112-160-224-315-450-630)		
Swivel			90 degrees towards wheel and 30 degrees away from wheel		
Morse taper			3		

Work-piece material

The present study is carried out with mild steel (AISI 1040). All the specimens are in the form of round bars of diameter 48mm and length 50mm.

5.5.2 Results and discussion

The generated surfaces are measured using the Talysurf instrument and further processed to get fractal dimension (D). The experimental results are used for further analysis using artificial neural networks to model fractal dimension. For ANN, full factorial design of experiments is considered and the design matrix and the experimental results of cylindrical grinding of mild steel work-pieces are presented in Table 5.13 (Barman and Sahoo, 2011).

To construct the multilayer feed-forward neural network models, three cutting parameters, namely work-piece speed, longitudinal feed and radial infeed are used as the input neurons and corresponding fractal dimension as the output neuron. Considering the full factorial design, a total of

Table 5.13 **Design matrix of process variables and the experimental results in grinding**

Std order	Run order	*N* Work-piece speed (rpm)	*f* Longitudinal feed (mm/rev)	*d* Radial infeed (mm)	*D*
1	22	1	1	1	1.46
2	45	1	1	2	1.48
3	7	1	1	3	1.40
4	37	1	1	4	1.46
5	54	1	2	1	1.47
6	38	1	2	2	1.44
7	26	1	2	3	1.45
8	42	1	2	4	1.43
9	13	1	3	1	1.47
10	43	1	3	2	1.43
11	63	1	3	3	1.42
12	59	1	3	4	1.42
13	32	1	4	1	1.41
14	5	1	4	2	1.45
15	40	1	4	3	1.44
16	34	1	4	4	1.45
17	51	2	1	1	1.49
18	10	2	1	2	1.47
19	14	2	1	3	1.45
20	21	2	1	4	1.46
21	1	2	2	1	1.43
22	64	2	2	2	1.42
23	48	2	2	3	1.44
24	61	2	2	4	1.41
25	23	2	3	1	1.47
26	31	2	3	2	1.45
27	53	2	3	3	1.35
28	29	2	3	4	1.39
29	12	2	4	1	1.45
30	56	2	4	2	1.45
31	2	2	4	3	1.43

(*Continued*)

Table 5.13 *Continued*

Std order	Run order	*N* Work-piece speed (rpm)	*f* Longitudinal feed (mm/rev)	*d* Radial infeed (mm)	*D*
32	46	2	4	4	1.44
33	25	3	1	1	1.47
34	3	3	1	2	1.45
35	19	3	1	3	1.46
36	33	3	1	4	1.48
37	8	3	2	1	1.48
38	49	3	2	2	1.47
39	17	3	2	3	1.44
40	6	3	2	4	1.47
41	18	3	3	1	1.42
42	44	3	3	2	1.47
43	15	3	3	3	1.43
44	57	3	3	4	1.45
45	36	3	4	1	1.43
46	35	3	4	2	1.39
47	28	3	4	3	1.45
48	41	3	4	4	1.42
49	11	4	1	1	1.45
50	47	4	1	2	1.48
51	30	4	1	3	1.44
52	27	4	1	4	1.47
53	39	4	2	1	1.47
54	16	4	2	2	1.45
55	9	4	2	3	1.46
56	52	4	2	4	1.47
57	24	4	3	1	1.47
58	50	4	3	2	1.41
59	4	4	3	3	1.46
60	20	4	3	4	1.42
61	62	4	4	1	1.45
62	60	4	4	2	1.45
63	58	4	4	3	1.45
64	55	4	4	4	1.44

64 sets of data points are obtained. These datasets are randomly chosen for training, testing and validation sets to get rid of any bias. Forty datasets for training, 12 datasets for testing and 12 datasets for validation are used. The data points are normalised so that the datasets range in –1 to +1. To construct the models, the hyperbolic tangent sigmoid function in the hidden layer and linear activation function in the output layer are considered. To train the network, different training algorithms, namely Levenberg-Marquardt (L-M) algorithm, gradient descent with variable learning rate and momentum (GDX) algorithm, scaled conjugate gradient (SCG) algorithm and Broyden-Fletcher-Goldfarb-Shanno (BFGS) quasi-Newton algorithm, are used. The Nguyen-Widrow weight initialisation algorithm has been applied. For generalisation of the networks, 'early stopping' technique is implemented. Here, the number of neurons in the hidden layer is decided by trial with increasing number of neurons. The neurons in the hidden layers are varied from 4 to 10 to find out the best neural network model for different training algorithms. In this study minimum mean squared error (MSE) is used as the selection parameter. Besides this parameter, for the network, correlation coefficient (R) and maximum absolute percentage error for the training dataset and testing dataset are calculated. MATLAB 6.5 is used to develop the neural network, train and test the network.

Performances of different networks trained with L-M, GDX, SCG and BFGS algorithms are presented in Table 5.14. It is seen that architecture 3-14-1 trained using L-M algorithm has the minimum mean squared error and the architecture is selected as the best performing network. The architecture also provides the maximum correlation coefficient (R) and minimum absolute percentage error. When the networks are trained using GDX algorithm, the

Table 5.14 Performance of different networks in grinding

Model	MSE values for training pattern				Maximum absolute % error			
	L-M	GDX	SCG	BFGS	L-M	GDX	SCG	BFGS
3-4-1	0.00031747	0.00043601	0.00043602	0.00046331	3.09	3.91	5.65	5.29
3-5-1	0.00028831	0.00042798	0.00033945	0.00044921	3.77	5.03	5.48	3.68
3-6-1	0.00025583	0.00038895	0.00028100	0.00043148	2.87	3.76	2.90	5.00
3-7-1	0.00026813	0.00031426	0.00024024	0.00038720	2.98	4.26	2.79	4.52
3-8-1	0.00029587	0.00025016	0.00035257	0.00031179	4.05	2.67	4.24	2.98
3-9-1	0.00023197	0.00035786	0.00039936	0.00023069	3.17	3.56	4.85	2.45
3-10-1	0.00017423	0.00035875	0.00030399	0.00027278	2.18	4.19	2.98	3.56
3-11-1	0.00017593	0.00025038	0.00026615	0.00035372	2.22	2.44	3.41	3.85
3-12-1	0.00013387	0.00025984	**0.00023966**	0.00033276	2.37	4.01	2.75	4.12
3-13-1	0.00023015	0.00025652	0.00028970	0.00026561	2.90	3.01	2.72	3.33
3-14-1	**0.00011341**	**0.00019131**	0.00029367	**0.00015634**	2.07	2.60	3.23	2.10
3-15-1	0.00011722	0.00027430	0.00036520	0.00021506	2.08	3.34	3.36	2.67
3-16-1	0.00022875	0.00024497	0.00031061	0.00024094	2.92	3.55	4.30	2.80

architecture 3-14-1 gives the minimum MSE and is selected as the best network. Architecture 3-12-1 provides the minimum MSE when the network is trained with SCG algorithm. So, architecture 3-12-1 is selected as the best network for training network using SCG algorithm. Again, it is observed that 3-14-1 architecture provides the optimum performance having minimum mean square error when the network is trained using BFGS quasi-Newton algorithm and subsequently the network is chosen as the best network.

Table 5.15 presents the comparative study of the performance among the best networks when trained with different training algorithms. Architecture 3-14-1 trained with L-M algorithm is selected as the best network based on the minimum MSE. It is also seen that the architecture has the minimum training error (2.07%). It implies that the experimental value and ANN predicted values lie very close to each other. It is observed from the regression analysis that correlation coefficient (R) is 0.914 for training pattern (Figure 5.12) and it is a good indication of the correlation between predicted and experimental outputs. The developed network is tested with 12 new patterns which are unknown to the developed network. The correlation coefficient for testing pattern is 0.816 (Figure 5.13). Comparative study of the experimental *D* and ANN predicted *D* is presented in Figure 5.14 and it is seen that ANN predicted and experimental fractal dimension are very close to each other. The mean absolute percentage error for testing pattern is 0.97%, which means the network predicts the fractal dimension with high accuracy. From the results, it is obvious that the network gives a good prediction of fractal dimension in grinding of mild steel and the network has generalisation capability.

Table 5.15 Comparison of network performances based on different training algorithms in grinding

Algorithm	Architecture	MSE for training	MSE for testing	Correlation coefficient (R)	Maximum training error (%)
L-M	**3-14-1**	**0.00011341**	**0.00033693**	**0.914**	**2.07**
GDX	3-14-1	0.00019131	0.00042063	0.823	2.60
SCG	3-12-1	0.00023966	0.00043855	0.832	2.75
BFGS	3-14-1	0.00015634	0.00038227	0.782	2.10

Figure 5.12 Performance of ANN model using regression analysis for training pattern in grinding

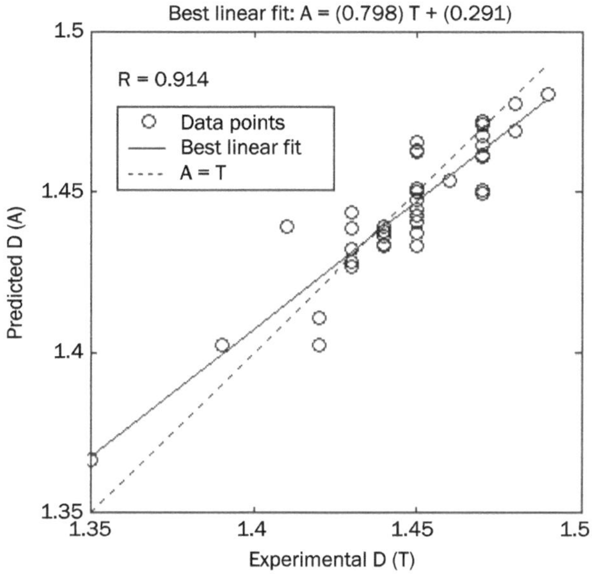

Figure 5.13 Performance of ANN model using regression analysis for testing pattern in grinding

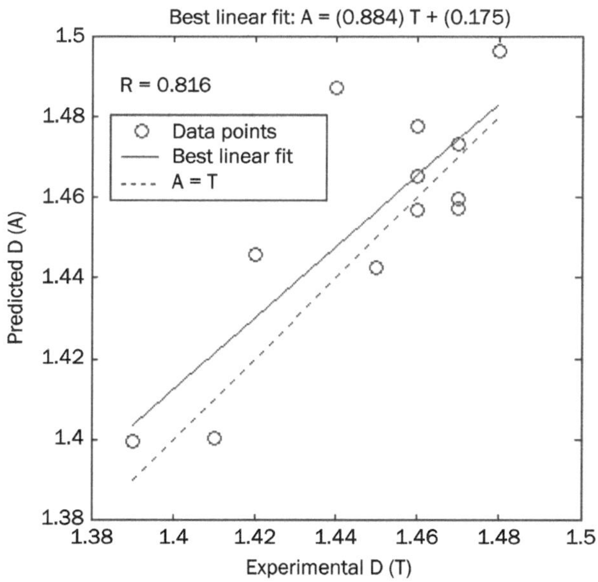

Figure 5.14 **Comparative study of experimental *D* and ANN predicted *D* for testing pattern using L-M algorithm in grinding**

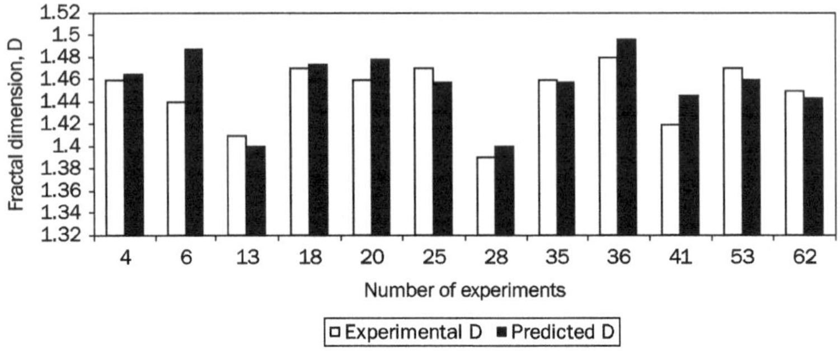

5.6 Electrical discharge machining

Electrical discharge machining (EDM) is a widespread machining technique used for all types of conductive materials including metals, metallic alloys, graphite, composites and ceramic materials. It is a non-conventional machining process used for machining of difficult-to-machine materials and shapes with a high degree of accuracy (El-Hofy, 2005). It is based on removing material from a part by means of a series of repeated electrical discharges created by electric pulse generated at short intervals between two electrodes; a tool electrode and a work-piece electrode. The electrodes are separated by a dielectric fluid that makes it possible to flush eroded particles from the gap between the electrodes. The electric spark raises the surface temperature of both the tool and work-piece to a point that is in excess of the melting or even boiling points of the substances. Thus material is mainly removed in the liquid and vapour phases, and the surface generated consists of debris that has either been melted or vaporised during machining. Since the tool does not physically contact the work-piece, no mechanical stress is exerted on the work-piece and the characteristics of the EDM process are thus not governed by the mechanical properties of the work-piece material. Instead, the thermal and electrical properties play a significant role in the process performance. The EDM performance is characterised by three parameters, namely material removal rate (MRR), electrode wear rate (EWR) and surface roughness. In this study, surface roughness is modelled based on fractal dimension for mild steel (AISI 1040) in EDM using ANN. The experimental details and the results are presented below.

5.6.1 Experimental details

Design of experiments

There are a large number of factors that can be considered for machining of a particular material in EDM. It is very difficult to conduct the experiment considering all the process variables. However, the review of literature shows that the following three machining parameters are the most widespread among the researchers and machinists to control the EDM process: pulse current (I, amp), pulse-on time (t_i, μs) and pulse-off time (t_o, μs). In the present study these are selected as design factors while other parameters have been assumed to be constant over the experimental domain. A face-centred central composite (FCC) design is used with three levels of each of the three design factors. Considering three factors and six replicates at the centre point, the present design contains 20 experiments, which were performed in a random order. The upper and lower limits of a factor are coded as +1 and −1 respectively, the coded value being calculated using Equation 5.8. The process variables with their values on different levels are listed in Table 5.16. The selection of the values of the variables is limited by the capacity of the machine used in the experimentation as well as the recommended specifications for different work-piece tool material combinations. Table 5.17 shows

Table 5.16 Variable levels used in EDM

Levels	Current	Pulse-on time	Pulse-off time
	(*I*, *amp*)	(t_i, μs)	(t_o, μs)
–1	3.125	50	50
0	6.250	100	75
1	9.375	150	100

Table 5.17 Design matrix of the FCC design in EDM

		Coded value		
Std. order	Run order	Current (*I*)	Pulse-on time (t_i)	Pulse-off time (t_0)
1	5	–1	–1	–1
2	2	1	–1	–1
3	8	–1	1	–1
4	12	1	1	–1
5	18	–1	–1	1
6	16	1	–1	1
7	14	–1	1	1
8	1	1	1	1
9	9	–1	0	0
10	11	1	0	0
11	6	0	–1	0
12	13	0	1	0
13	19	0	0	–1
14	3	0	0	1
15	7	0	0	0
16	20	0	0	0
17	10	0	0	0
18	15	0	0	0
19	17	0	0	0
20	4	0	0	0

the experimental matrix of the FCC design employed in the present study.

Machine used

The machine used for carrying out the machining operations is a 'Toolcraft A25' EDM machine having the stepped drive servo system and filtration flushing capability. It is capable of generating maximum pulse current of 25 ampere, pulse-on time of 2,000 μs and pulse-off time of 2,000 μs. The specification of the machine is presented in Table 5.18.

Table 5.18 **Specification of the EDM equipment**

Particulars	Specification
Trade name	TOOL CRAFT A 25
Type of construction	'C' type
Work table	300mm × 200mm
Fixed work tank	465mm × 270mm × 200mm
Table longitudinal movement	100mm
Table cross movement	175mm
Maximum dielectric level over table	140m
Maximum work-piece height	90mm
Maximum work-piece weight	45kg
Servo Head	
Servo system	Stepped drive
Quill travel	150mm
Electrode platen size	100mm sq.
Accuracy of quill movement	0.01mm over 200mm
Dielectric system	
Filtration flushing	better than 10 microns
Flushing	side, 1.23litres/min. (max)
Flushing pressure	15kPa
Generator	
Models	A 25

(*Continued*)

Table 5.18 *Continued*

Working current	25 A max. through current selector
Pulse-on time setting	2 to 2,000 microseconds
Pulse-off time setting	2 to 2,000 microseconds
Power source connection	400/440v, 50Hz, 3-ph supply

Work-piece material

The present study was carried out with AISI 1040 mild steel. All the specimens were in the form of 20mm x 20mm x 4mm blocks.

Tool electrode used

Electrolytic copper having 99.9% copper in composition and density 8,904kg/m^3 was used as tool electrode since it worked better in combination with the work-piece material considered in the present study. The tool electrode was in the form of a cylinder of diameter 15.9mm and 50mm in length mounted axially in line with the work-piece. The tool electrode was given negative polarity whereas the work-piece has positive polarity (Puertas et al., 2005). The properties of the tool electrode are given in Table 5.19. Kerosene was used

Table 5.19 **Electrode material properties**

Particulars	Specifications
Material	Electrolytic copper
Composition	99.09% copper
Density	8,904kg/mm^3
Melting point	1,083^{0}C
Conductivity	101.41% IACS
Tensile strength	23.47kg/mm^2

as dielectric because of its high flashpoint, good dielectric strength, transparent characteristics and low viscosity and specific gravity.

5.6.2 Results and discussion

The generated machined surfaces are measured with Talysurf and further processed to calculate fractal dimension. Experimental results of fractal dimension are presented in Table 5.20 (Barman and Sahoo, 2010).

Table 5.20 Experimental results in EDM

Std. order	Run order	D
1	5	1.413
2	2	1.310
3	8	1.330
4	12	1.276
5	18	1.426
6	16	1.306
7	14	1.426
8	1	1.283
9	9	1.333
10	11	1.286
11	6	1.346
12	13	1.343
13	19	1.346
14	3	1.316
15	7	1.356
16	20	1.306
17	10	1.363
18	15	1.306
19	17	1.330
20	4	1.316

Using these experimental data, artificial neural network models for predicting fractal dimension have been developed using multi-layer feed-forward back propagation algorithm.

To construct and train the network, pulse current intensity (I), pulse-on time (t_i) and pulse-off time (t_o) are used as the input parameters and corresponding fractal dimension (D) is used as the output. Only one hidden layer is used in the networks. The numbers of neurons are varied to select best network based on the minimum mean squared error (MSE). For training of the network, Levenberg-Marquardt algorithm is used. All the observations of the FCC design are used in the training of the networks.

The MSE values for different architectures are presented in Table 5.21. From the table, it is seen that for mild steel, increase in the number of neurons in the hidden layer beyond four does not improve in the performance and thus the network with 3-4-1 architecture is selected based on minimum mean squared error. For mild steel, the maximum absolute percentage error is obtained as 2.42%, which implies that the ANN model outputs and experimental outputs are very close to each other. The comparative study of experimental and ANN model predicted fractal dimension is presented in Figure 5.15. From this figure also, it is clear that the predicted and experimental

Table 5.21 MSE data for different architecture in EDM

Serial No.	ANN Architecture	MSE
1	3-3-1	0.00020499
2	3-4-1	0.00015456
3	3-5-1	0.00015456
4	3-6-1	0.00015456

Figure 5.15 Comparative study of experimental *D* with ANN model predicted *D* in EDM

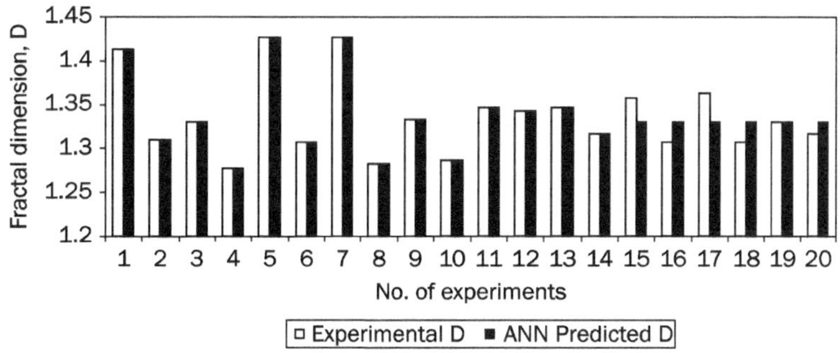

fractal dimension are close to each other. The performance of the neural network is measured by carrying out the regression analysis. It is observed from the regression analysis that correlation coefficient (R) is 0.957 (Figure 5.16) and it

Figure 5.16 Performance of ANN model using regression analysis in EDM

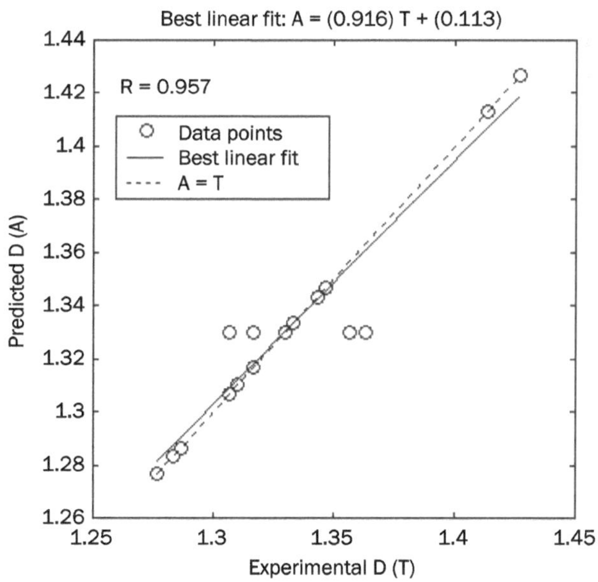

is a good indication of the correlation between predicted and experimental outputs.

5.7 Conclusion

Using back-propagation algorithm, multilayer artificial neural networks are developed for predicting fractal dimension (*D*) for different machining operations, namely CNC milling, CNC turning, cylindrical grinding and EDM. For all the machining operations, work-piece material is chosen as mild steel (AISI 1040). The generated machined surfaces are measured and processed to find fractal dimension. To develop the neural network models, fractal dimension is used as the output of the network whereas machining parameters are chosen as the input neurons to the network. For CNC milling, CNC turning and EDM processes, Levenberg-Marquardt training algorithm is used to train the network. However, for cylindrical grinding operations, four different training algorithms are used to train the networks and compared, but the networks trained with Levenberg-Marquardt training algorithm give the best performance. For all machining operations, it is very clear from the results that developed ANN models can predict fractal dimension very accurately. From the regression analyses, also, it is clear that the experimental and predicted fractal dimensions are in good agreement.

5.8 References

Barman, T.K. and Sahoo, P. (2009) Artificial neural network modelling of fractal dimension in CNC turning and comparison with response surface model, *Journal of Machining and Forming Technologies*, 1(3–4): 197–220.

Barman, T.K. and Sahoo, P. (2010) ANN modeling of fractal dimension in electrical discharge machining and comparison with RSM models, *Journal of Modern Manufacturing Technology*, 2(2): 105–18.

Barman, T.K. and Sahoo, P. (2011) ANN modeling of fractal dimension in grinding of mild steel and comparison with response surface model, *Journal of Mechatronics and Intelligent Manufacturing*, 2(1–2): 41–52.

Benardos, P.G. and Vosniakos, G.C. (2003) Predicting surface roughness in machining: a review, *International Journal of Machine Tool & Manufacture*, 43(8): 833–44.

Berry, M.V. and Lewis, Z.V. (1980) On the Weierstrass-Mandelbrot fractal function, *Proceedings of Royal Society, London*, A 370: 459–84.

Bhushan, B., Wyant, J.C. and Meiling, J. (1988) A new three-dimensional non-contact digital optical profiler, *Wear*, 122: 301–12.

Bigerelle, M., Najjar, D. and Iost, A. (2005) Multiscale functional analysis of wear: a fractal model of the grinding process, *Wear*, 258: 232–9.

Brown, C.A. and Savary, G. (1991) Describing ground surface texture using contact profilometry and fractal analysis, *Wear*, 141: 211–26.

Demuth, H. and Beale, M. (1998) *Neural network toolbox user's guide*, version 3.0. Natick, MA: The Mathworks Inc.

El-Hofy, H.A.G. (2005) *Advanced machining processes*. New York: McGraw-Hill.

Foresee, F.D. and Hagan, M. (1997) Gauss-Newton approximation to Bayesian learning, *Proceedings of International Joint Conference on Neural Networks*, 3: 1930–5.

Ge, S. and Chen, G. (1999) Fractal prediction models of sliding wear during the running-in process, *Wear*, 231: 249–55.

Han, J.H., Ping, S. and Shengsun, H. (2005) Fractal characterization and simulation of surface profiles of copper electrodes and aluminum sheets, *Materials Science and Engineering A*, 403: 174–81.

Hasegawa, M., Liu, J., Okuda, K. and Nunobiki, M. (1996) Calculation of the fractal dimensions of machined surface profiles, *Wear*, 192: 40–5.

Haykin, S. (2007) *Neural networks: a comprehensive foundation*, 2nd edition. New Delhi: Prentice-Hall.

He, L. and Zhu, J. (1997) The fractal character of processed metal surfaces, *Wear*, 208: 17–24.

ISO 4287:1997 (1997) *Geometrical Product Specification (GPS) – Surface Texture: Profile Method – Terms, Definitions and Surface Texture Parameters*. Geneva: International Organization of Standardization.

Jahn, R. and Truckenbrodt, H. (2004) A simple fractal analysis method of the surface roughness, *Journal of Materials Processing Technology*, 145: 40–5.

Jiang, Z., Wang, H. and Fei, B. (2001) Research into the application of fractal geometry in characterizing machined surfaces, *International Journal of Machine Tools and Manufacturing*, 41: 2179–85.

Kang, M.C., Kim, J.S. and Kim, K.H. (2005) Fractal dimension analysis of machined surface depending on coated tool wear, *Surface and Coatings Technology*, 193(1–3): 259–65.

Ling, F.F. (1990) Fractals, engineering surfaces and tribology, *Wear*, 136: 141–56.

MacKay, D.J.C. (1992) Bayesian interpolation, *Neural Computation*, 4(3): 415–47.

Majumdar, A. and Bhushan, B. (1990) Role of fractal geometry in roughness characterization and contact mechanics of surfaces, *Transactions ASME: Journal of Tribology*, 112: 205–16.

Majumdar, A. and Tien, C.L. (1990) Fractal characterization and simulation of rough surfaces, *Wear*, 136: 313–27.

Mandelbrot, B.B. (1967) How long is the coast of Britain? Statistical self-similarity and fractional dimension, *Science*, 156: 636–8.

Mandelbrot, B.B. (1982) *The Fractal Geometry of Nature*. New York: W.H. Freeman.

Nayak, P.R. (1971) Random process model of rough surfaces, *Transactions ASME: Journal of Lubrication Technology*, 93: 398–407.

Puertas, I., Luis, C.J. and Villa, G. (2005) Spacing roughness parameters study on the EDM of silicon carbide, *Journal of Materials Processing Technology*, 164–165: 1590–6.

Sahoo, P. (2005) *Engineering Tribology*. New Delhi: Prentice-Hall.

Sahoo, P. and Ghosh, N. (2007) Finite element contact analysis of fractal surfaces, *Journal of Physics D: Applied Physics*, 40: 4245–52.

Sahoo, P., Barman, T.K. and Routara, B.C. (2008) Fractal dimension modeling of surface profile and optimization in CNC end milling using response surface method, *International Journal of Manufacturing Research*, 3(3): 360–77.

Sayles, R.S. and Thomas, T.R. (1978) Surface topography as a non-stationary random process, *Nature*, 271: 431–4.

Thomas, T.R. (1982) Defining the microtopography of surfaces in thermal contact, *Wear*, 79: 73–82.

Tricot, C., Ferlans, P. and Baran, G. (1994) Fractal analysis of worn surfaces, *Wear*, 172: 127–33.

Venkatesh, K., Bobji, M.S. and Biswas, S.K. (1998) Some features of surface topographical power spectra generated by conventional machining of a ductile metal, *Materials Science and Engineering A*, 252: 153–5.

Venkatesh, K., Bobji, M.S., Gargi, R. and Biswas, S.K. (1999) Genesis of workpiece roughness generated in surface grinding and polishing of metals, *Wear*, 225–229: 215–26.

Whitehouse, D.J. (1982) The parameter rash, is there a cure? *Wear*, 83: 75–8.

Yan, W. and Komvopoulos, K. (1998) Contact analysis of elastic-plastic fractal surfaces, *Journal of Applied Physics*, 84(7): 3617–24.

Zhang, Y., Luo, Y., Wang, J.F. and Li, Z. (2001) Research on the fractal of surface topography of grinding, *International Journal of Machine Tools and Manufacture*, 41: 2045–9.

Zhu, H., Ge, S., Huang, X., Zhang, D. and Liu, J. (2003) Experimental study on the characterization of worn surface topography with characteristic roughness parameter, *Wear*, 255: 309–14.

6

Predicting forces and damage in drilling of polymer composites: soft computing techniques

Inderdeep Singh, Pawan Kumar Rakesh and Jagannath Malik

Abstract: Polymer composites are finding substantial application in the most diverse engineering fields. Drilling in composites is an unavoidable operation which leads to substantial damage around the drilled hole. Conventional regression techniques have been used to develop models for drilling forces and drilling-induced damage. The present trend is towards the use of soft computing techniques for development of predictive models. This chapter briefly reviews the state-of-the-art concepts of soft computing techniques, namely artificial neural network (ANN), genetic algorithm (GA) and particle swarm optimisation (PSO). The application of these techniques for developing optimised, generic predictive models for drilling forces and drilling-induced damage has also been highlighted.

Key words: drilling; polymer composites; soft computing; artificial neural network.

6.1 Drilling of polymer composites

Polymer matrix composites (PMCs) are composed of matrix and reinforcement. The primary function of the matrix is to transfer stresses between the reinforcing fibres and to protect the fibres from mechanical as well as environmental damage. The fibres impart their special mechanical and physical property to enhance the matrix properties.

Drilling is the most important machining operation practised widely to ensure the assembly operations of the complex PMC products. A summary of research publications in the field of drilling of PMC laminates is presented in Figure 6.1. The percentage contribution in terms of number of papers published in the last 20 years has been used as the data. It has been observed that 25 per cent of the total papers published in the field have been in the last three years. Therefore, it can be stated with confidence that the area of drilling of PMCs has generated substantial interest among the research fraternity. The main focus has been to minimise the drilling forces (thrust forces and torque) and the resulting drilling-induced damage.

In order to reduce or avoid drilling-induced damage, a number of approaches have been attempted. The optimisation of the various drilling parameters such as the drill point geometry, the drill material and the operating variables (cutting speed and the feed rate) has been the focus of researchers. There have been attempts to correlate the output characteristics, such as the thrust force and the torque, with the drilling-induced damage, but still the exact theory of damage-generation and quantification eludes the composites family.

Figure 6.1 Number of papers published in the field of drilling of PMCs in last two decades

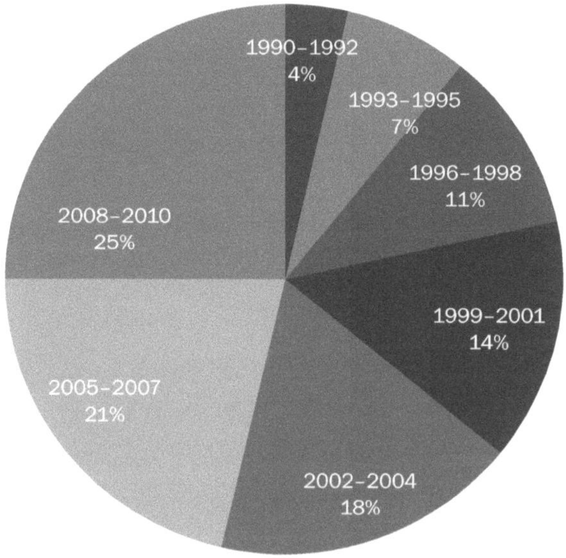

The published literature in the field of drilling of PMCs has been carefully analysed and it has been found that the drill point geometry, the drill material and the feed rate are the most significant input parameters to be optimised (Figure 6.2a). Another important point to note is that the focus is on recording the drilling forces and correlating the forces with the drilling-induced damage. There have been attempts to quantify the drilling-induced damage in the form of delamination (Figure 6.2b). The summary of focus areas among the various input parameters (cutting speed, feed rate, drill material and drill geometries) and the output parameters (drill wear, surface roughness, drilling forces, delamination and mechanical behaviour of composite laminates with drilled hole) in the published data is shown in Figure 6.2.

Figure 6.2 **Summary of focus areas in the field of drilling of PMCs**

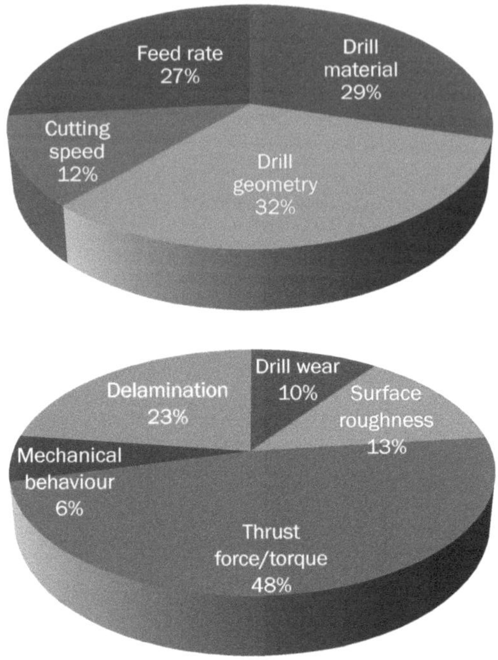

6.1.1 *Input parameters*

Operating variables

The cutting speed and the feed rate are the two most important operating variables in the drilling process. Increasing the feed rate leads to increased drilling forces and increased drilling-induced damage, whereas increasing the cutting speed leads to an increased drill wear and minimum drilling-induced damage. The cutting speed and feed rate influence the thrust force and torque (Mohan et al., 2005). The drilling forces are directly correlated with the drilling-induced damage. The drilling-induced damage in composite laminates results from the cutting action of the drill point and the feed rate is the most critical parameter that influences the damage (Caprino and Tagliaferri, 1995). Higher cutting speeds increase the risk of thermal damage, as it causes the softening of the matrix material. A reduction in feed rate has a consequence on temperature build-up, increasing the risk of thermal damage (Lin and Chen, 1999). The high-speed cutting plays a major role in drilling of polymer composites and reducing damage at the entry and exit side of the drilled hole (Krishnamoorthy et al., 2009).

The various significant elements that influence the drilling process of PMCs are shown in Figure 6.3. The operating parameters, drill parameters and material parameters significantly affect the drilling process. The decisions in term of optimal selection of the drill point geometry and drill material as well as the cutting speed and feed rate are important to ascertain the damage-free drilling of PMCs.

Drill point geometry

The drill point geometry is important in terms of making a damage-free hole in PMCs. The optimisation of drill

Figure 6.3 Elements of the drilling system in context of PMCs

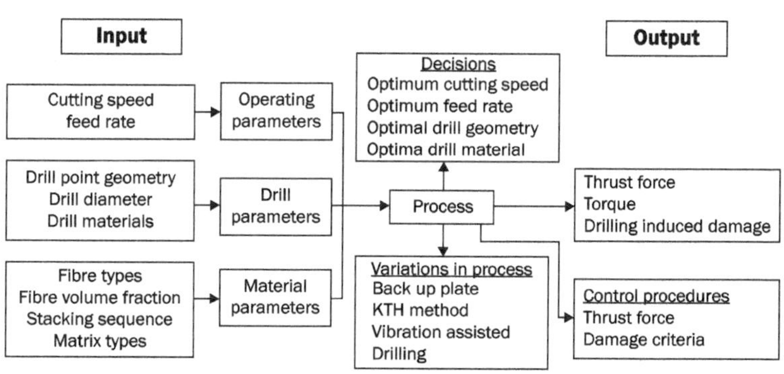

point geometries is the major focus area in order to minimise the drilling forces and drilling-induced damage. Optimised drill point geometry has resulted in better-quality holes and reduction in the drilling forces. The drilling of PMCs with 17 drill bit configurations was conducted and it was found that drill geometry influences the drilling-induced damage. The 8-facet drill point appears to achieve its success by combining the free cutting and the other qualities of a split point twist drill with a long taper angle at the shoulder, which tends to minimise fibre breakout. The drill point ensured the pre-loading of the fibres under tensile stress before being cut by the shearing action (Miller et al., 2007). The drilling has been carried out with different drill point geometries (namely, twist drill, saw drill, candle stick drill, core drill and step drill) and the thrust force varies with the drill geometry as well as with the feed rate (Hocheng and Tsao, 2006). The thrust force behaviour of distinct geometries (stub length, brad and spur) was noted, while drilling glass fibre reinforced plastic (GFRP) composites. The thrust force increased with the feed rate; however, minimum thrust force was recorded while using the brad and spur drill. The brad and spur drill resulted in better performance as the specific cutting pressure and the thrust force were lower than the stub length. Moreover, the delamination and the surface roughness were smaller for brad and spur than the stub length (Davim et al., 2007). A hollow grinding drill was developed for drilling of PMCs. It resulted in a minimum thrust force and better hole quality as compared with the twist drill (Jain and Yang, 1994). The four different types of drill point geometries (Figure 6.4) were used for drilling of GFRP composites. The drilling forces were minimum for jo drill but the drilling-induced damage was minimum

Figure 6.4 Different types of drill geometry

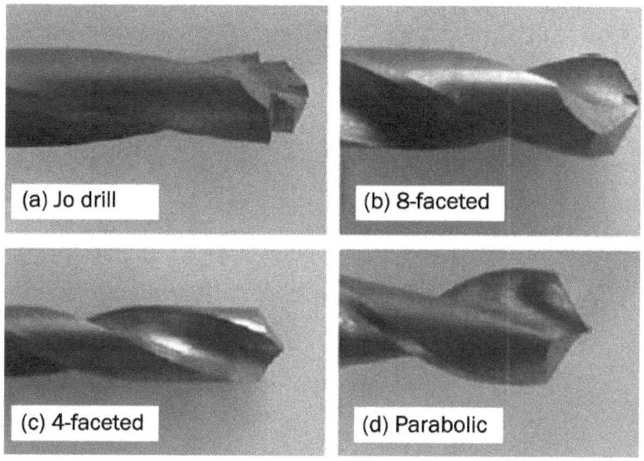

for 8-facet drill. Jo drill resulted in minimum torque as compared with 4-facet and 8-facet drill (Singh and Bhatnagar, 2006). The drilling-induced damage was dependent on the drilling parameters. The drill point angle influences the drilling-induced damage in the composites (Palanikumar et al., 2008). The different types of drill geometries have been used for drilling of polymer composites such as twist drill, jo drill, 8-facet drill, trepanning tool, parabolic drill, brad and spur, saw drill, candle stick drill, core drill, step drill, C-shaped drill, dagger drill, conical drill, raconw drill.

Drill material

The judicious selection of the drill material is very important in drilling of PMCs. The reinforcement is harder to cut and the drill wears out quickly. Three different materials – high speed steel (HSS), high speed steel-cobalt (HSS-Co), and carbide drill – were used for drilling of composites. Increasing cutting speed leads to increased drill wear, larger entrance and exit burrs, larger damage area and decreased number of holes drilled. The flank and crater wear takes place on HSS and HSS-Co drills, whereas minor flank wear was noticed on carbide drills. Carbide drills have the highest hot hardness, produced the lower thrust force, smallest burrs, negligible damage area and the highest hole quality (Ramulu et al., 2001). The effect of drill geometry, coating and drilling conditions was evaluated while drilling in carbon fibre reinforced plastic (CFRP) composites and it was reported that drill life increased significantly when uncoated tungsten carbide stepped drills were used at higher feed rate. On drilling by HSS and HSS-Co drills, increased feed rate resulted in increasing the thrust force and torque, drilling-induced damage along thickness direction and increased surface roughness. It is an established fact that HSS drills should be avoided for drilling of PMCs as greater number of holes to failure are drilled with carbide drills with highest hole quality. The helical flute cemented carbide (K10) drill point gives better performance than the helical flute HSS drill (Shyha et al., 2009). The performance of three different types of carbide drills, either uncoated or TiN or diamond like carbon (DLC) coating was studied. The results indicated that coated drills did not reduce drill wear or drilling-induced damage when drilling CFRP composites. However, in comparison with HSS drills, the carbide drills with or without coatings presented superior wear resistance (Murphy et al., 2002). Cemented diamond drills resulted in better surface quality as compared with plated diamond and plated solid drills, while drilling CFRP composites (Quan and Zhong, 2009). The wear in the case of HSS drills was larger than that for carbide drills. Drill wear at the centre of cutting edges increased with the number of drilled holes (Sakuma et al., 1984).

6.1.2 Output characteristics

Drilling forces

As shown in Figure 6.2, the major emphasis of the research efforts worldwide has been towards recording of the drilling forces. The behaviour of drilling forces has been analysed and efforts have been made to correlate the forces with the resulting damage around the drilled hole. The thrust force has been found to be lower at lower feed rates. The thrust force is a highly sensitive carrier of information about the status of the drilling process. The large oscillations of the thrust force were observed while drilling kevlar fibre reinforced plastics (KFRP), which might be attributed to the in-homogeneity inside the lamina and to the presence of interfaces in between the laminate. These oscillations can be interpreted as non-uniform distribution of the thrust force along the drill cutting edge and the poor inter-laminar strength of the laminates that can cause piercing effects at the interfaces (Ilio et al., 1991). The thrust force increases with feed rate due to the fact that the shear area is increased (Abrao et al., 2008). The effect of the drill size on thrust force and torque was observed and it was noted that thrust force and torque increased with drill diameter, due to the increase in the shear area (Ei-Sonbaty et al., 2004). Increasing the chisel edge length of the drill geometry results in an increase in the thrust force. A minimum thrust force was recorded while drilling with eccentric twist or an eccentric candle stick drill. The thrust force has been established as the root cause of damage in drilling of PMCs (Fujii et al., 1971). The trepanning tool reduces the thrust force and torque during drilling of GFRP composites as compared with the twist drill. The torque behaviour was similarly influenced by the drill point geometries and the feed rate. A decrease in the feed rate reduces the thrust force, which can reduce the extent of induced delamination (Mathew et al., 1999). The four different drill point geometries (namely, 4-facet drill, 8-facet drill, parabolic drill and jo drill) have been used for drilling of PMCs. The drilling forces were recorded using the drill dynamometer (Kistler, type 9272). It has been seen that the drill point geometry affects the thrust force at the same operating variables (Figure 6.5).

Figure 6.5 Thrust force for different drill point geometries

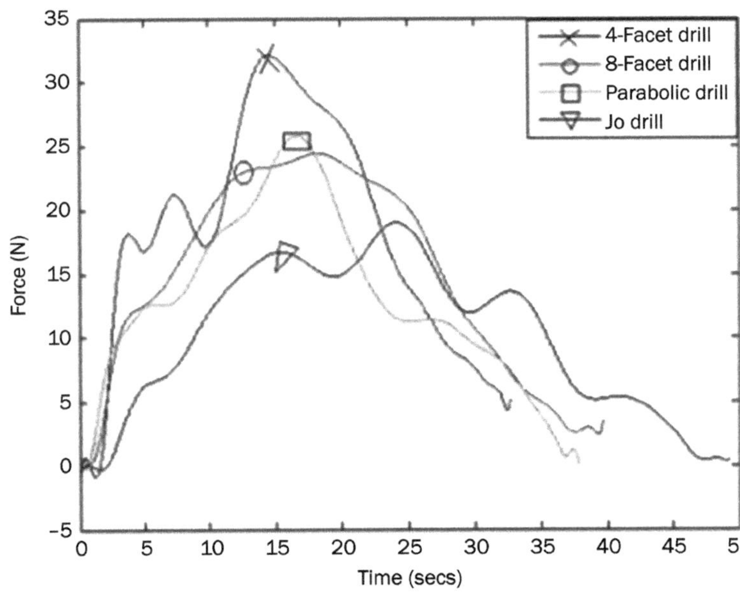

Drilling-induced damage

Delamination is one of the most common damage forms associated with drilling of polymer composites. The most common types of drilling-induced damage mechanisms are peel-up delamination at entrance and push-down delamination at exit side of the drilled hole. The various other types of drilling-induced damage found are fibre pull-out, matrix burning, chipping, matrix cracking, debonding, circularity defect, fuzzing and spalling. The three significant damage mechanisms which have been identified that cause the growth of delamination are the plate bulge, crack opening and the fibre breaking. The drilling-induced damage not only causes a poor surface finish and quality but also results in the performance deterioration of the composite product over a period of time (Di-Paolo et al., 1996). The drilling-induced damage is strongly dependent on the feed rate. When the feed rates are high, the failure modes show the signs of the impact damage, with step-like delamination, intra-laminar cracks and the high-density micro failure zones. If the feed rate is sufficiently low, the failure consists essentially of delamination mainly originating near the intersection between the conical surface generated by the main cutting edges and the cylindrical surface of the drilled hole (Caprino and Tagliaferri, 1995).

Efforts have been undertaken for the quantification of the damage around the drilled hole using a number of methodologies. Ultrasonic C-scan, acoustic emission, radiography, computerised tomography (CT), digital image analysis, damage quantification using microscope and laser-based imaging have been the most prominent techniques adopted to quantify the drilling-induced damage in PMCs. The drilling-induced damage area has also been quantified using the non-destructive dye penetrant test. The drilling-induced damage in PMCs while drilling with a 4-facet drill is shown in Figure 6.6. During drilling in unidirectional GFRP composites, the damage zone around the drilled hole is elliptical in shape with the major axis parallel to the direction of the fibres.

A brief summary of the research work done in the field of drilling of PMCs has been discussed in the previous sections. As is quite clear from Figure 6.3, the drilling system in the

Figure 6.6 Drilling-induced damage in unidirectional GFRP laminates

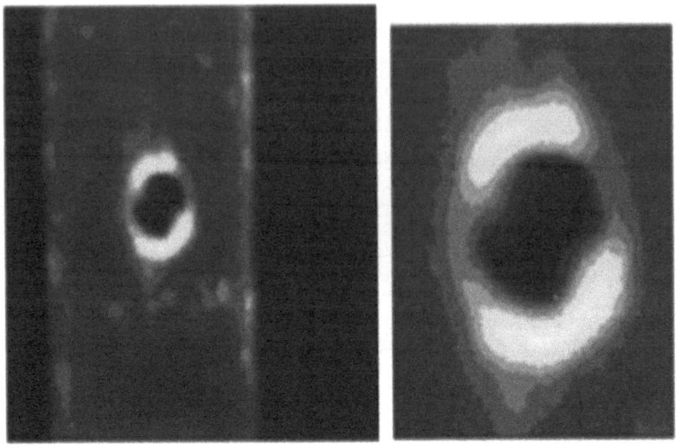

context of PMCs is quite complex and a judicious selection of a combination of input parameters is required for avoiding or completely eliminating drilling-induced damage. It is difficult to develop generic models for different types of PMCs. Therefore empirical or semi-empirical models developed for predicting the drilling forces as well as the drilling-induced damage cannot be generalised.

It is clear from Figure 6.7 that most of the studies made so far are experimental in nature, with only 9 per cent resulting in development of predictive models for the various output characteristics. The research efforts in the field of simulation of the drilling behaviour of PMCs are also not encouraging. Although theoretical models have been proposed for calculating the critical thrust force beyond which drilling-induced damage may occur, such models have their own limitations. As the relation between the various input and output parameters is quite complex, there have been attempts worldwide to use soft computing techniques to model the drilling behaviour of PMCs. Soft computing techniques can be used to predict the drilling forces and drilling-induced damage in drilling of PMCs.

Figure 6.7 **Research directions in the field of drilling of PMCs**

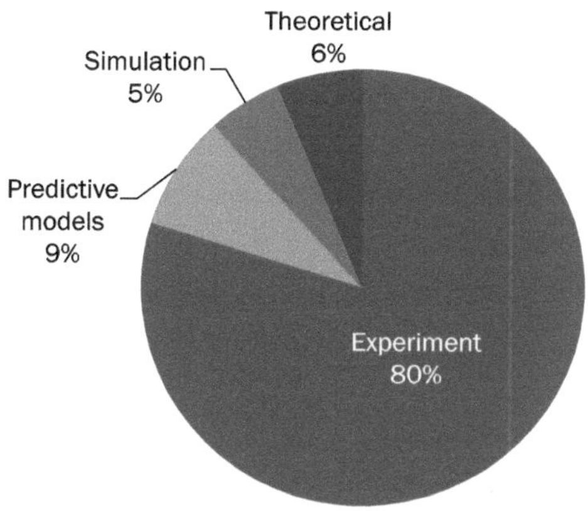

The following sections highlight the application of soft computing techniques in the field of drilling of PMCs.

6.2 Soft computing techniques

Soft computing is a collection of techniques spanning many fields that fall under various categories in computational intelligence. The major divisions of soft computing are: neural networks, evolutionary computation and fuzzy logic. A number of other soft computing techniques, namely Bayesian networks, support vector machines (SVM) and neuro-fuzzy systems, also have popular application in various disciplines. Hybrid algorithms, wavelet theory, chaos theory, to name a few, supplement the list.

Soft computing uses soft techniques blended with classical artificial neural computing and intelligence, which is capable of learning from statistical data, modelling and analysing very complex phenomena for which conventional methods have limitations. The applicability of conventional computing is governed and bound by a computer science concept called

NP-complete, which means there is a proportionality between the size of a problem and the amount of resources, e.g. memory and computation time, needed to solve the problem. Soft computing helps to overcome NP-complete constraint by using inexact methods (biological or nature inspired) to give useful near-correct solutions.

Opposite to 'hard computing', soft computing is rarely prescriptive in its solution to a problem. Each and every possible solution is not programmed. Instead, the problem is represented in such a way that the 'state' of the system can be measured and compared with some desired state repeatedly, which is the basis for adapting/updating the system's parameters. With time (iteration), the system slowly converges towards the desired solution. This is the basic approach employed by most of the soft computing techniques such as genetic algorithms, particle swarm optimisation and neural networks.

Soft computing techniques exploit and use the given tolerance of imprecision in data, partial truth and uncertainty to come up with a near accurate solution for a particular problem. Inductive reasoning plays an important role in soft computing. Major advantages of soft computing include its robustness and high tolerance in noisy input environments. Lotfi Zadeh, founder of fuzzy logic, says of computing that 'the major imperatives of computing with words are that computing with words is a necessity when the available information is too imprecise to justify the use of numbers and when there is tolerance in imprecision which can be exploited to achieve robustness and low cost solution' (Zadeh, 1994).

Soft computing became a formal computer science area of study in the early 1990s. Conventional computational approaches precisely analyse relatively simple systems with mathematical modelling and formulae. With the advancement in science, demand for analysing more complex systems arising in biology, medicine, the humanities, management sciences, and similar fields falls in the limitations of conventional mathematical and analytical methods. It should be pointed out that simplicity and complexity of systems are relative and many conventional mathematical or analytical models have been both challenging and very productive in solving these.

6.2.1 Application of soft computing in drilling of polymer composites

As has been discussed in previous sections, the drilling of PMCs is a complex phenomenon. The effect of various input parameters and the effect of a combination of input parameters on the output response characteristics are not well understood. Therefore, to suggest the behaviour of forces and the amount of damage on the basis of change in the input parameters is a difficult task.

From Figure 6.8, the complex relationship between the drilling input and output variables can be observed. The nature of the problem to predict output characteristics from input variables is fairly complex. It is advisable to use soft computing tools such as back propagation algorithm and hybrid models of ANN with other optimisation techniques in order to improve the prediction accuracy.

Figure 6.8 Drilling process with input and output variables

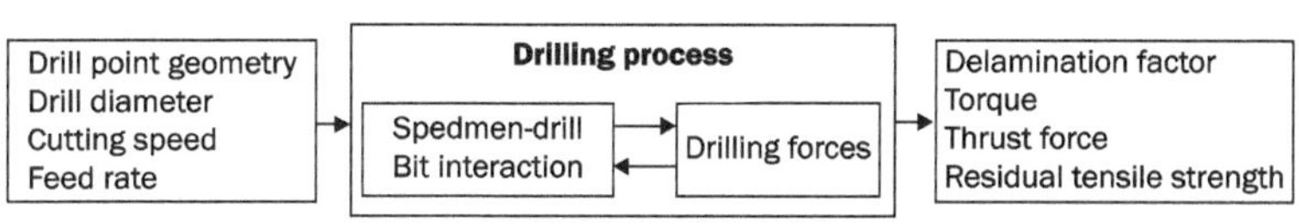

6.2.2 Artificial neural networks and information processing

Neural networks (NNs) are collections of interconnected neurons similar to those found in biological brains. Artificial neurons are crude approximations of the neurons found in brains. ANNs are networks of artificial neurons, and hence constitute crude approximations to parts of real brains (Bullinaria, 2005). These may be simulated on conventional computers or realisable physical devices (analogue or digital). From a practical point of view, an ANN is just a massive parallel computational system consisting of many processing units (neurons) connected together in a specific way in order to perform a particular task. In a more general sense it can be said: 'An ANN is an information processing paradigm that is inspired by the way biological nervous systems, such as the brain, process information. It is composed of a number of interconnected processing elements (neurons) working in unison to solve specific problems' (Stergiou and Siganos, 2011).

ANN can be used as a modelling tool to approximate any system when the input-output characteristic is available (Ham and Kostanic, 2002). In supervised learning NN establishes a probable relationship between input and output. The statistical input-output mapping information is distributed over the entire synaptic connections among neurons constituting the ANN, so that the response of this black box is equivalent to the actual system. General topology of an artificial NN is shown in Figure 6.9.

The most common topology of ANN is a feed-forward multi-layer network where neurons are arranged in layers, from input to output with interconnecting weights between two adjacent layers. Each neuron within the network processes one or more inputs coming into it and produces an output (decided by activation function) for the next layer. At each

Figure 6.9 Topology of a feed-forward neural network

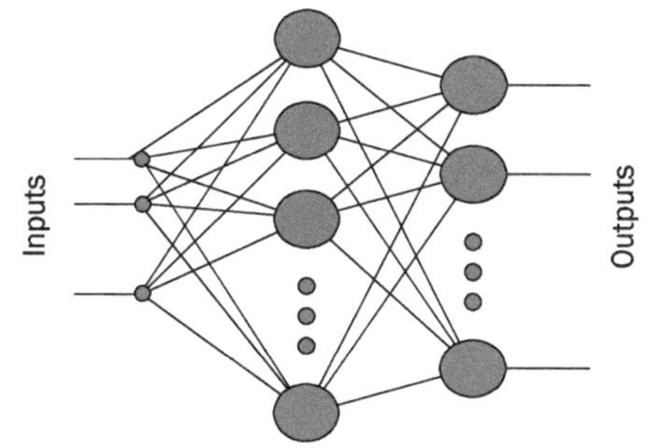

neuron, every input has an associated weight which modifies (scale) the strength of each input value. A mathematical model of a single neuron is shown in Figure 6.10.

A single hidden layer network model with a sufficient number of hidden neurons is enough to approximate any physical model. There is no universally accepted and mathematically proven rule to decide for the number of hidden neurons in a hidden layer. Generally, it is advised to start with fewer hidden neurons and go for higher numbers incrementally. Care and compromise should be taken between computational time and accuracy.

Figure 6.10 Structure of a single neuron

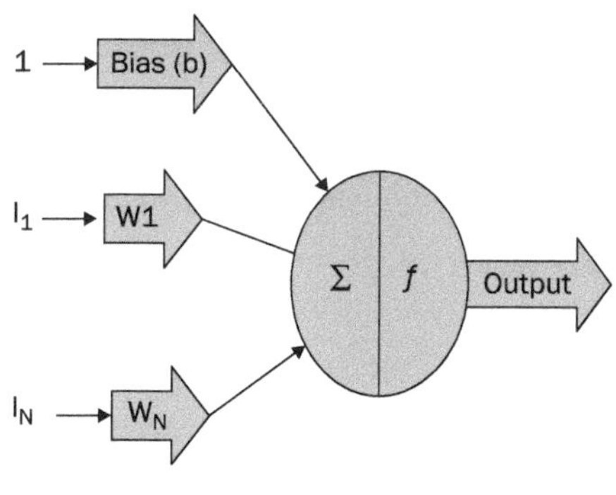

$$\text{Output} = f\{\Sigma(I_i + W_i) + b\}$$

Whether our NN is a simple single hidden layer network, or a much more complicated multi-layer network with special activation functions, a systematic procedure (training algorithm) for determining appropriate connection weights has been developed (Haykin, 1999). The general procedure is to have the network learn the optimum weights from a representative set of training data. Trained networks are expected to generalise, i.e. respond appropriately with input data they were not trained on. The famous error back-propagation learning, the gradient descent method, is usually used to train the network.

In regard to the data set, the prediction accuracy of a well-trained ANN is fairly unaltered (same accuracy level) for very large data sets as well as fewer data points. Prediction accuracy and robustness of any ANN depend on how well it is trained. Also the prediction accuracy of ANNs greatly depends on the range of variables in which it is being trained. Accuracy outside the training span is greatly decreased. A large data set with continuous uniform step variation for all input variables and corresponding actual output is always desirable for both training and testing purposes. But for experiments involving huge costs of operation, the generation of a larger data set is not economically viable. For the same reason, a limited yet sufficient data set with more than one level for each input variable should be generated.

Finally, it can be concluded that ANNs can represent non-linear input-output mapping as a network pattern, and the manner in which the information is distributed over synaptic connections is a matter of no great importance.

ANN in drilling of polymer composites

An NN based on back-propagation (BP) algorithm with two hidden layers was developed for modelling of delamination factor in drilling of GFRP composites. It was concluded that the well-trained BP network model can precisely predict the delamination in drilling of GFRP composites (Latha and Senthilkumar, 2010).

The prediction and evaluation of thrust force and surface roughness in drilling of composite using a candle stick drill was attempted using Taguchi method and the ANN. The objective was to establish a correlation between the feed rate, cutting speed and drill diameter with the thrust force and surface roughness. The correlations were obtained by multi-variable regression analysis and radial basis function network (RBFN) and compared with the experimental results. In the confirmation tests, RBFN demonstrated more effectiveness than regression analysis. The obtained results with regression show the maximum deviation of 14 per cent. On the other hand, the errors of thrust force and surface roughness with RBFN were below 2 and 4 per cent, respectively (Tsao and Hocheng, 2008).

An NN approach has been proposed for predicting delamination in drilling of GFRP. A software system has been developed and tested for performing an online prediction of the damage both at the entry and exit side of the tool. The prediction is made possible by means of feed-forward NNs, which are able to give a 'measure' of the damage process as a function of the cutting conditions. Two architectures of NNs have been proposed and investigated: the former is able to sort the predicted delamination in four categories (no damage, low, medium and high damage); the latter is able to directly predict an average value of the damage (Dinia, 2003).

An experimental approach for the prediction of thrust force produced by a step drill using linear regression analysis and RBFN has been proposed. In the confirmation tests, RBFN (errors within 0.3 per cent) has been shown to be a better predictive model than multi-variable linear regression analysis (errors within 28 per cent) (Tsao, 2008).

A model was developed using both regression and neuro-fuzzy models to predict thrust force and torque in the drilling operation of GFRP composites. The neuro-fuzzy model showed much better accuracy than the regression model. The average absolute errors for thrust force and torque were found to be 5.83 per cent and 4.57 per cent respectively with the neuro-fuzzy model, whereas for the regression model, they were 17.3 per cent and 26.67 per cent respectively (Jayabal and Natarajan, 2010).

A feed-forward back propagation ANN system was used to investigate the influence of input parameters on the thrust force and torque in the drilling process. A multi-layer perceptron model was constructed using the input parameters of cutting speed, feed rate and volume fraction of the reinforced plastics. Output parameters were the thrust force and torque. The predicted forces based on the ANN model were found to be in very good agreement with the unexposed experimental data. The prediction accuracy was 93.24 and 94.17 per cent for thrust force and torque, respectively (Hayajneh et al., 2009).

The ANN topology was adopted as a predictive tool. The cutting speed, feed rate, drill diameter and drill point geometry were used as the input parameters. The drilling-induced damage was the output. The maximum absolute error for training patterns was found to be 12.7 per cent and the minimum was 0.1 per cent, and for most cases the error was less than 5 per cent. The maximum absolute error for testing patterns was found to be 13.19 per cent and the minimum was 0.35 per cent (Mishra et al., 2010).

An ANN predictive approach was developed with inputs as the drill point geometry, the feed rate and the cutting speed with residual tensile strength of composite with drilled holes as the output. The maximum absolute error for testing patterns was found to be 9.39 per cent, while for most cases the error was less than 5 per cent (Mishra et al., 2010).

6.2.3 Genetic algorithms

Genetic algorithms (GA) are adaptive algorithms for finding the global optimum solution for an optimisation problem. The GA is characterised by binary representation of individual solutions, simple problem independent crossover and mutation operators and a proportional selection rule (Holland, 1975).

The GA is a parallel search heuristic that works on the process of natural biological evolution. GA is a widely used tool for optimisation and fast search applications. On being inspired from biological evolution, its optimisation process is modelled as genetic processes such as chromosomes, selection, crossover and mutation. GA optimises input parameters for a fitness function by modelling the search space by population of strings (also known as genome type of chromosomes). In traditional GA techniques, chromosomes are represented as a binary string of 0s and 1s with equivalent decimal numbers to be used as input parameters to fitness function. Therefore each chromosome corresponds to a fitness value in the search space of fitness function. GA makes use of multiple chromosomes and searches to make parallel search in fitness function. It evolves new generation of chromosomes by processes of crossover and mutation to introduce variety in chromosome population. GA prefers chromosomes with high fitness value over those with low fitness value in the chromosome pairing (crossover) and hence generates random combination to search for better population (having higher fitness values). The process of mutation reduces the probability of deadlock in population evolution. To understand the GA concept, consider the standard procedure of the GA as outlined in Figure 6.11.

Figure 6.11 Outline of genetic algorithm

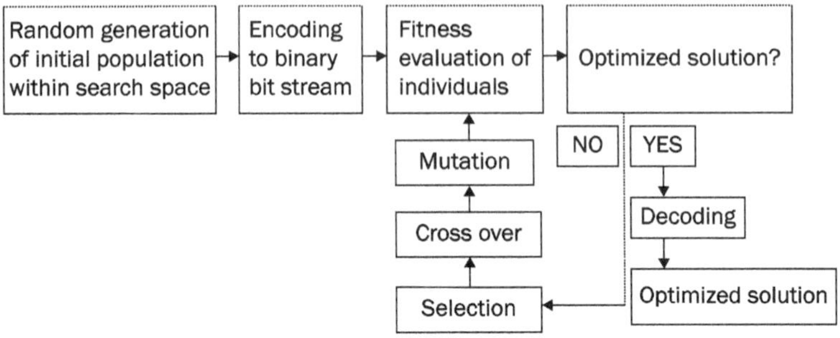

Hybrid GA-ANN model

Keeping in mind the advantages and limitations of back propagation NN and GA, many researchers have tried successful hybridisation of both algorithms. Combining the inherent philosophy and properties of both algorithms, a hybrid model can overcome their individual limitations.

GA can be used to:

- generate the weights of an NN;
- generate the architecture of an NN;
- generate both the architecture and weights of an NN.

An adaptive genetic algorithm optimised artificial neural network (AGA-ANN) topology has been adopted for predicting damage. The experimental data for drilling of GFRP composites were used for training and testing the model.

Parallel GA models are being used to heuristically search for NN weights. All the weights are encoded with chromosomes, each represented as an N-bit binary string. Also, two modifications in GA to make it adaptive in operation and hence enabling it to perform more purposeful search have been introduced. On each of the iterations, selected individuals participate in the process of reproduction (crossover and mutation) and form new generation. The population selection was done with the help of the 'roulette wheel technique', such that the probability of any individual getting selected is directly proportional to its relative fitness value among its corresponding population.

If *N* is the number of hidden layers in an NN model working on a normalised 4-dimentional feature input vector and single output, the total number of weights to be optimised becomes 4×N+N = 5N. Therefore the proposed method makes use of 5N mutually independent GA models each having *Z* individuals. Each individual is a search site in weight space and encodes a weight valued between –X to +X (*X* being 2) with a *P* (*P* being 10) binary string (chromosome). On each of the GA iterations, the proposed method evaluates 5N × 10GA individuals for a stability criterion named as fitness values.

Each individual pair in refined population undergoes crossover with a probability equal to crossover rate (which is 0.75 in the present case). It is worth noticing that exchange of high significance bit (HSB) segment causes more change in an individual's equivalent weight than the low significant bit (LSB) segments. Therefore smooth and steady convergence of GA is proposed. In early iterations of GA, the crossover segments should be picked from HSBs and the position of segment should gradually shift towards LSBs governed by Newman Gaussian probability density function (NG-PDF) with adaptive mean and fixed Gaussian width with iterations. As the position of crossover segment shifts from HSB to LSB based on accuracy and iteration count criterion, the GA search type gets changed from coarse to fine and hence stabilises faster than traditional GA. To break the deadlock condition, mutation site is also selected by NG-PDF similar to crossover position and mutation rate of 0.02.

Three predictive models, namely BP-ANN, GA-ANN and AGA-ANN, were developed in MATLAB. Network parameters, i.e. adaptive learning rate for BP-ANN, crossover and mutation probability for GA-ANN, adaptive crossover site and mutation probability for proposed AGA-ANN, were chosen optimally. For each input pattern, the predicted value of the output variable (the ratio of damage area to hole area) has been compared with the respective experimental value for different networks in Figure 6.12a. The network was tested with the 20 test data points that had not been used for the purpose of training. From the test results, it was observed that the predicted values are very close and follow almost the same trend as the experimental values for AGA-ANN

Figure 6.12 (a) Experimental and predicted output for different models on test data; (b) Histogram plot of prediction accuracy for AGA-ANN

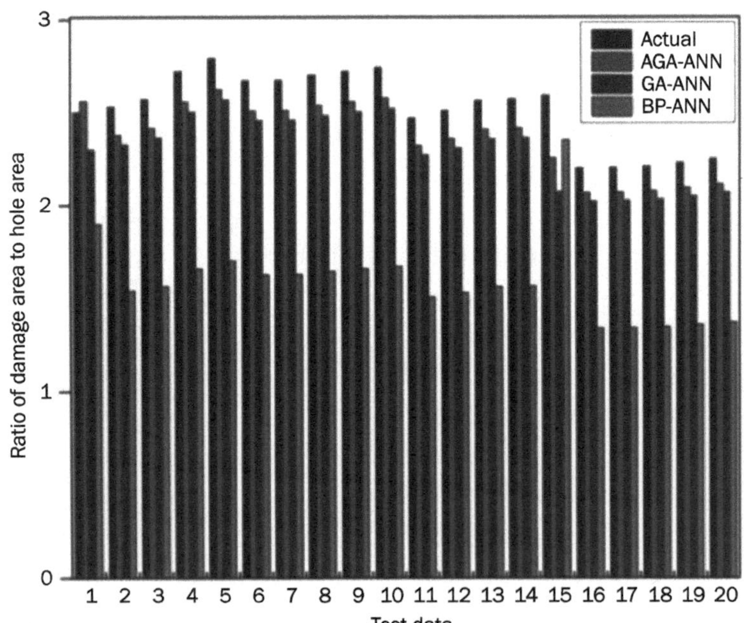

(a)

(*Continued*)

Figure 6.12 *Continued*

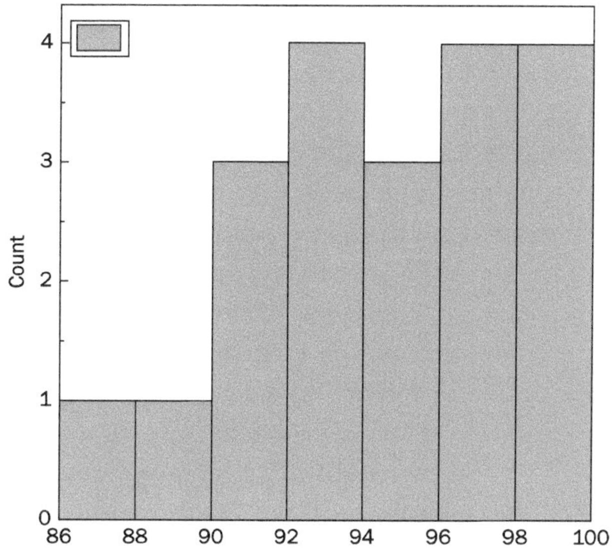

(b)

network. The mean prediction accuracy over all test data was 94.43 per cent. The histogram plot of prediction accuracy for AGA-ANN model is shown in Figure 6.12b.

From the above results it can be concluded that the proposed AGA-ANN model shows good prediction accuracy (the mean prediction accuracy was ≈ 94 per cent on test data) compared with the other two models. This may be a useful tool for calculating the drilling-induced damage for various combinations of input parameters. Prior knowledge of the potential damage could help engineers to select the optimum process parameters as well as the drill point geometry for making damage-free holes in GFRP composites.

6.2.4 *Particle swarm optimisation*

Particle swarm optimisation is a population-based stochastic optimisation technique (Eberhart and Kennedy, 1995), inspired by social behaviour (intelligence) of swarm. The PSO algorithm is the outcome of a theory called 'social dynamics theory'. It says: 'The individuals searching for solutions learn from the experiences of others and individuals that interact frequently become similar. Culture affects the performance of individuals that comprise it.' Social behaviour (interaction) increases the ability of an individual and 'intelligence arises from interactions among individuals'.

The advantage of PSO is that it is comparatively simple in concept, easy to construct (code and implement) and offers high computational performance. This is an efficient algorithm capable of solving difficult multi-dimensional and multi-objective optimisation problems in various disciplines. Its problem-solving capability has already been exploited and applied successfully to solve many real-life problems involving complex and constrained optimisation.

A problem is given, and a way to evaluate a proposed solution to it exists in the form of a fitness function. Then a population of individuals made by random guesses to the problem solution is initialised. These individuals are candidate solutions, also known as the particles, hence the name particle swarm. Each particle has a very simple 'memory' of its personal best solution so far, called 'p_{best}'. The global best solution for each iteration is also found and labelled 'g_{best}'. It is the best value, obtained so far by any particle in the population. Each particle makes this information available to their neighbours, analogous to social interaction. Once set in motion each particle is moved a certain distance from its current location, influenced by a random amount by the 'p_{best}' and 'g_{best}' values to improve fitness. During this iterative process, the particles gradually settle down to an optimum solution. In a minimisation optimisation problem, problems are formulated so that 'best' simply means the position with the smallest objective value.

Combined evolutionary search concept of PSO and gradient-based training of ANN has been used to develop a predictive model and optimise inter-connecting synaptic weights for estimating the drilling-induced damage in CFRP laminates. Using the well-balanced search (exploration and exploitation) property of hybrid PSO-ANN, the developed model showed satisfactory performance. The results of the model were in good agreement with the training and the testing data. The plots of target and predicted output values (normalised) for testing data for PSO-ANN model are given in Figure 6.13. The results were quite satisfactory, with less than 10 per cent error in 73 per cent of cases for PSO-ANN model.

Figure 6.13 Plot of ANN-PSO output (normalised) against testing data

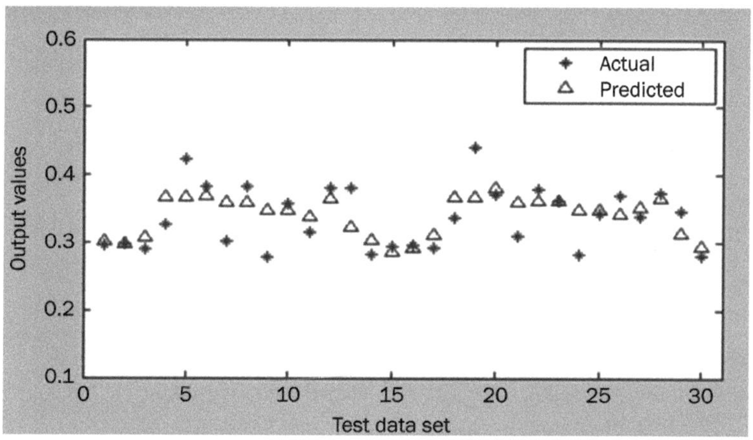

6.2.5 Inverse model development

The previous section dealt with application of soft computing techniques for developing predicting models for drilling forces and drilling-induced damage in PMCs. This section presents a brief discussion on how to develop an inverse model problem to optimise the drilling variables to achieve a pre-specified (target) level for forces or delamination factor. The objective can be achieved in two ways:

- To train and develop a predictive network model using either of the algorithms discussed in previous sections with the input and output domain reversed. Here the main problem is the dimensions of input/output variables. Sometimes the mapping of lower dimension data to higher dimension is not so efficient, resulting in poor generalisation of the system. The developed predictive model lacks interpolation capability and establishing proper statistical information between input-output variables of drilling.
- The second method is to use combined heuristic and generalisation properties of hybrid GA-ANN or PSO-ANN models. Here a neural model is developed in the same manner discussed previously with input/output unaltered. Thereafter, a PSO or GA is adopted to search for the optimal combinations of input variables to achieve target output with the cost function being trained NN.

Suppose experimental data of input parameters and any of the output parameters for a number of drilling experiments are available. ANN model with input/output parameters can be designed and trained. With proper design parameters – i.e. number of hidden layers, number of neurons in each layer, learning rate and training algorithm – a robust predictive model can be developed with good generalisation and heuristics that approximate the real drilling model. For a specific drill bit with drill diameter (both belong to developed experimental data), in order to optimise cutting speed and feed rate for either damage-free drilling or to have specified thrust force and torque value, a possible solution is to generate random initial combinations for both speed and feed rate within allowable limits. Thereafter, either PSO or GA can be adopted to find the best ones. For this, trained NN serves as the cost function.

6.3 References

Abrao, A.M., Rubio, J.C., Faria, P.E. and Davim, J.P. (2008) The effect of cutting tool geometry on thrust force and delamination when drilling glass fiber reinforced plastic composite, *Mater Des*, 29: 508–13.

Bhatnagar, N., Naik, N.K. and Ramakrishnan, N. (1993) Experimental investigations of drilling of CFRP composites, *Mater Manuf Processes*, 8: 683–701.

Bullinaria, J.A. (2005) Available from: *http://www.cs.bham.ac.uk/~jxb/IAI/w3.pdf* [accessed 10 April 2011].

Caprino, G. and Tagliaferri, V. (1995) Damage development in drilling glass fiber reinforced plastics, *Int J Mach Tools Manuf*, 35: 817–29.

Davim, J.P., Rubio, J.C. and Abrao, A.M. (2007) A novel approach based on digital image analysis to evaluate the delamination factor after drilling composite laminates, *Compos Sci Technol*, 67: 1939–45.

Dinia, G. (2003) On-line prediction of delamination in drilling of GFRP by using a neural network approach, *Mach Sci Technol*, 7: 295–314.

Di-Paolo, G., Kapoor, S.G. and Devor, R.E. (1996) An experimental investigation of the crack growth phenomenon for drilling of fiber reinforced composite materials, *J Eng Ind Trans ASME*, 118: 104–10.

Eberhart, R.C. and Kennedy, J. (1995) A new optimizer using particle swarm theory, *Proceedings of the Sixth International Symposium on Micro Machine and Human*

Science, Nagoya, Japan, Piscataway, NJ: IEEE Service Center, 39–43.

Ei-Sonbaty, I., Khashaba, U.A. and Machaly, T. (2004) Factors affecting the machinability of GFR/epoxy composites, *Compos Struct*, 63: 329–38.

Fujii, S., Devries, M.F. and Wu, S.M. (1971) An analysis of the chisel edge and the effect of the d-theta relationship on drill point geometry, *J Eng Ind Trans ASME*, 93: 1093–105.

Ham, F.M. and Kostanic, I. (2002) *Principles of neuro computing for science and engineering*. New York: Tata McGraw-Hill.

Hayajneh, M.T., Hassan, A.M. and Mayyas, A.T. (2009) Artificial neural network modeling of the drilling process of self-lubricated aluminum/alumina/graphite hybrid composites synthesized by powder metallurgy technique, *J Alloys Compd*, 478: 559–65.

Haykin, S. (1999) *Neural networks: a comprehensive foundation*, 2nd edition. Upper Saddle River, NJ: Prentice Hall.

Hocheng, H. and Tsao, C.C. (2006) Effects of special drill bits on drilling-induced delamination of composite materials, *Int J Mach Tools Manuf*, 46: 1403–16.

Holland, J.H. (1975) *Adaptation in natural and artificial systems*. Ann Arbor: University of Michigan Press.

Ilio, A.D., Tagliaferri, V. and Veniali, F. (1991) Cutting mechanisms in drilling of aramid composites, *Int J Mach Tools Manuf*, 31: 155–65.

Jain, S. and Yang, D.C.H. (1994) Delamination-free drilling of composite laminates, *J Eng Ind Trans ASME*, 116: 475–81.

Jayabal, S. and Natarajan, U. (2010) Regression and neuro fuzzy models for prediction of thrust force and torque in drilling of glass fiber reinforced composites, *J Sci Ind Res*, 69: 741–5.

Krishnamoorthy, A., Boopathy, S.R. and Palanikumar, K. (2009) Delamination analysis in drilling of CFRP composites using response surface methodology, *J Compos Mater*, 43: 2885–901.

Latha, B. and Senthilkumar, V.S. (2010) Application of artificial intelligence for the prediction of delamination in drilling GFRP composites, *Int J Precis Technol*, 1: 314–30.

Lin, S.C. and Chen, J. (1999) Drilling unidirectional glass fiber reinforced composite materials at high speeds, *J Compos Mater*, 33: 827–51.

Mathew, J., Ramakrishnan, N. and Naik, N.K. (1999) Investigations into the effect of geometry of trepanning tool on thrust and torque during drilling of GFRP composites. *J Mater Process Technol*, 91: 1–11.

Miller, S.F., Blau, P.J. and Shih, A.J. (2007) Tool wear in friction drilling, *Int J Mach Tools Manuf*, 47: 1636–45.

Mishra, R., Malik, J., Singh, I. and Davim, J.P. (2010) Neural network approach for estimating the residual tensile strength after drilling in uni-directional glass fiber reinforced plastic laminates, *Mater Des*, 31: 2790–5.

Mishra, R., Malik, J. and Singh, I. (2010) Prediction of drilling-induced damage in unidirectional glass fiber reinforced plastic laminates using an artificial neural network, *PI MECH ENG B-J ENG*, 224: 733–8.

Mohan, N.S., Ramachandra, A. and Kulkarni, S.M. (2005) Influence of process parameters on cutting force and torque during drilling of glass-fiber polyester reinforced composites, *Compos Struct*, 71: 407–13.

Murphy, C., Byrne, G. and Gilchrist, M.D. (2002) The performance of coated tungsten carbide drills when machining carbon fiber-reinforced epoxy composite materials, *PI MECH ENG B–J ENG*, 216: 143–52.

Palanikumar, K., Rubio, J.C., Abrao, A.M., Correia, A.E. and Davim, J.P. (2008) Influence of drill point angle in high speed drilling of glass fiber reinforced plastics, *J Compos Mater*, 42: 2585–97.

Quan, Y. and Zhong, W. (2009) Investigation drilling grinding of CFRP, *Front Mech Eng China*, 4: 60–3.

Ramulu, M., Branson, T. and Kim, D. (2001) A study on the drilling of composite and titanium stacks, *Compos Struct*, 54: 67–77.

Sakuma, K., Yokoo, Y. and Seto, M. (1984) Study on drilling of reinforced plastics (GFRP and CFRP) – relation between tool material and wear behavior, *Bull JSME*, 27: 1237–44.

Shyha, I.S., Aspinwall, D.K., Soo, S.L. and Bradley, S. (2009) Drill geometry and operating effects when cutting small diameter holes in CFRP, *Int J Mach Tools Manuf*, 49: 1008–14.

Singh, I. and Bhatnagar, N. (2006) Drilling of uni-directional glass fiber reinforced plastic (UD-GFRP) composite laminates, Int *J Adv manuf Technol*, 27: 870–6.

Stergiou, C. and Siganos, D. (2011) Available from: *http://www.doc.ic.ac.uk/~nd/surprise_96/journal/vol4/cs11/report.html* [accessed 10 April 2011].

Tsao, C.C. (2008) Prediction of thrust force of step drill in drilling composite material by Taguchi method and radial basis function network, *Int J Adv Manuf Technol*, 36: 11–18.

Tsao, C.C. and Hocheng, H. (2008) Evaluation of thrust force and surface roughness in drilling composite material using Taguchi analysis and neural network, *J Mater Process Technol*, 203: 342–8.

Zadeh, L.A. (1994) Fuzzy logic, neural networks, and soft computing, *Communications of the ACM*, 37: 77–84.

7

Minimising burr size in drilling: integrating response surface methodology with particle swarm optimisation

V.N. Gaitonde, S.R. Karnik and J. Paulo Davim

Abstract: The exit burr in drilling has several undesirable features with regard to product quality and functionality. Hence, selection of optimal process parameters for minimising burr size at production stage is essential to reduce the deburring cost and time. This chapter highlights the application of particle swarm optimisation (PSO) for burr size minimisation in drilling of AISI 316L stainless steel. The experiments were planned as per Box-Behnken design (BBD), and mathematical models for burr height and burr thickness were developed using response surface methodology (RSM) with cutting speed, feed, drill diameter and point angle as process parameters. The developed burr size models were then employed with PSO to determine the best combination of process parameters for a given drill diameter, which results in minimal burr height and burr thickness. The importance of larger point angle for bigger drill diameter values for minimising burr size is clearly evidenced in PSO simulation results.

Key words: drilling; burr size; Box-Behnken design of experiments; response surface methodology; particle swarm optimisation.

7.1 Introduction

Drilling is one of the most common operations broadly used in the manufacturing industries, including the aerospace and automotive sectors. The burr is a plastically deformed projected material generated on the part edge during drilling. The formation of the exit burrs has several undesirable features on quality and reliability of the product and also degrades the performance in precision components (Koelsch, 2001). These burrs can interfere with the assembly of parts and can cause jamming and misalignment, which in turn causes numerous problems for product functionality and hence requires considerable attention in production research. During machining, the exit burrs are injurious, as they hit the cutting edge and cause groove wear (Arai and Nakayama, 1986). On the other hand, the standard deburring tools are required for removing the burrs, which are formed inside a cavity. Gillespie (1996) reported that the deburring cost was 30 per cent of the total cost of the finished components and the edge finishing as well as secondary finishing operations are also found to be very difficult to automate. Even though secondary processes such as mechanical, thermal, electrochemical, ultrasonic deburring, spot facing, abrasive jet machining, laser deburring, mechanised cutting and hand filing have been performed to remove the exit burrs (Dornfeld, 1992), many of these processes are costly and hence there is a need to reduce the deburring effort.

Several researchers have studied the effects of various process parameters on burr formation during drilling of several work materials (Dornfeld et al., 1999; Ko and Lee, 2001; Lauderbaugh, 2009; Lin, 2002; Stein and Dornfeld, 1997). The finite element models (Guo and Dornfeld, 2000; Min et al., 2001a; Saunders, 2003), empirical drilling charts (Kim and Dornfeld, 2000; Kim et al., 2001; Min et al., 2001b), mathematical models based on response surface methodology (Gaitonde et al., 2009) have been developed for predicting and analysing the burr formation in drilling. Recently, modified Taguchi methods have also been employed for minimising the burr size in drilling (Gaitonde et al., 2007; Gaitonde and Karnik, 2007; Gaitonde et al., 2008; Hunag, 2004; Tosun, 2006).

Process modelling by response surface methodology (RSM) using statistical design of experiments (DOE) is proved to be an efficient modelling tool (Montgomery, 2003). The methodology not only reduces the cost and time, but also gives the required information about the direct and interaction effects of process parameters. On the other hand, the genetic algorithms (GA) and particle swarm optimisation (PSO) are the established multi-response optimisation tools. However, unlike GA, PSO does not have evolution operators, namely, crossover and mutation. In PSO, the potential solutions known as particles fly through the problem space by following the current optimum particles. The PSO is not only easy to implement but only a few parameters need to be adjusted as compared with GA (Elbeltagi et al., 2005; Konstantinos et al., 2004). Hence, PSO tool has been selected as an optimisation tool for the current study.

Due to greater corrosive resistance properties, AISI 316L stainless steel work material finds wide applications in aircraft properties, medical apparatus and chemical industries. However, burr formation during drilling is a major problem due to the highly ductile nature of the material. One of the most common tool materials for machining stainless steel work material is HSS (Kim et al., 2001). Hence, an attempt has been made in this research to determine the best combination of process parameters for minimising the burr size at exit of the holes during drilling of AISI 316L stainless steel using HSS twist drills. The effects of process parameters such as cutting speed, feed, drill diameter and point angle are analysed simultaneously on burr height and burr thickness using RSM-based mathematical models. The developed models are then employed with PSO for the selection of the optimal process parameters to simultaneously minimise burr height and burr thickness.

7.2 Response surface methodology

The response surface methodology (RSM) is a collection of mathematical and statistical techniques useful for constructing the models and analysing the problems in which several independent variables or controllable factors influence a dependent variable or response (Montgomery, 2003). In RSM, if all the independent variables are assumed to be measurable, then the response surface can be expressed as (Montgomery, 2003):

$$Y = \phi(x_1, x_2, x_3, \ldots\ldots, x_i, \ldots\ldots, x_k) \tag{7.1}$$

where, Y is the response of the system, ϕ is the response function, x_i is the independent variable i and k is the number of variables. If the mathematical form of response surface is not known, then it can be approximated by a polynomial within the experimental region. The higher the degree of polynomial, the better the correlation but at the higher costs of experimentation.

The RSM-based procedure consists of selection of proper design of experiments (DOE), development of an appropriate mathematical model of response surface with the best fittings of the experimental data and graphical representation of interaction effects of process parameters. The RSM has been applied for developing the mathematical models in the form of multiple regression equations. For the development of regression equations related to various quality characteristics of any process, the second order response surface has been assumed as (Montgomery, 2003):

$$Y = b_0 + \sum_{i=1}^{k} b_i X_i + \sum_{i=1}^{k} b_{ii} X_i^2 + \sum_{i=1}^{k-1} \sum_{j=i+1}^{k} X_i X_j \tag{7.2}$$

Equation 7.2 contains linear, squared and cross product terms of X_i's variables and the terms b_i, b_{ii} and b_{ij} are the regression coefficients, which are determined by (Montgomery, 2003):

$$b = (X^T X)^{-1} X^T Y \tag{7.3}$$

where b is the matrix of parameter estimates; X is the calculation matrix, which includes main and interaction terms; X^T is the transpose of X; and Y is the matrix of response. In order to determine the regression coefficients of second order mathematical models, the central composite rotatable and the Box-Behnken experimental designs are commonly used (Montgomery, 2003).

7.3 Particle swarm optimisation

The particle swarm optimisation (PSO) was developed by Kennedy and Eberhart, which is a relatively recent heuristic search method, whose mechanics are inspired by swarming or collaborative behaviour of biological populations. Each potential solution in the searching space is seen as a '*bird*', known as a '*particle*' (Kennedy and Eberhart, 2001) and all the particles have fitness values, which are evaluated by fitness function that is to be optimised, and have velocities, which direct the flying of particles. The birds fly through the problem space by following the current optimum particles. If the search space is D-dimensional, then the ith particle of the population, known as *swarm*, is represented by a D-dimensional vector $s = (s_1, s_2, \ldots, s_D)$. The velocity of this particle is represented by another D-dimensional vector $v = (v_1, v_2, \ldots, v_D)$. The best previously visited position of the ith particle is designated as $p_{best} = (p_1, p_2, \ldots, p_D)$. If g is the index of the best particle in the swarm (g_{best}) and the superscripts denote the iteration number, then the swarm is manipulated by the following equations:

$$v_i^{k+1} = w^k v_i^k + c_1 r_1^k (p_i^k - s_i^k) + c_2 r_2^k (p_g^k - s_i^k) \tag{7.4}$$

$$s_i^{k+1} = s_i^k + v_i^{k+1} \tag{7.5}$$

where w is the inertia weight; c_1 and c_2 are the cognitive and social parameters known as learning factors; r_1 and r_2 are the random numbers uniformly distributed in the range [0–1]; $i = 1, 2, \ldots N$, where N is the size of the swarm and $k = 1, 2, \ldots$ is the current iteration. The inertia weight regulates the trade-off between exploration and exploitation. A large weight facilitates global search, known as exploration, whereas, smaller weight tends to facilitate fine-tuning the current search area, known as exploitation. The fine-tuning of c_1 and c_2 parameters may result in faster convergence of the algorithm and alleviation of the local minima. On the other hand, r_1 and r_2 parameters are used

Figure 7.1 **Modification of a searching point concept by PSO in a two-dimensional space**

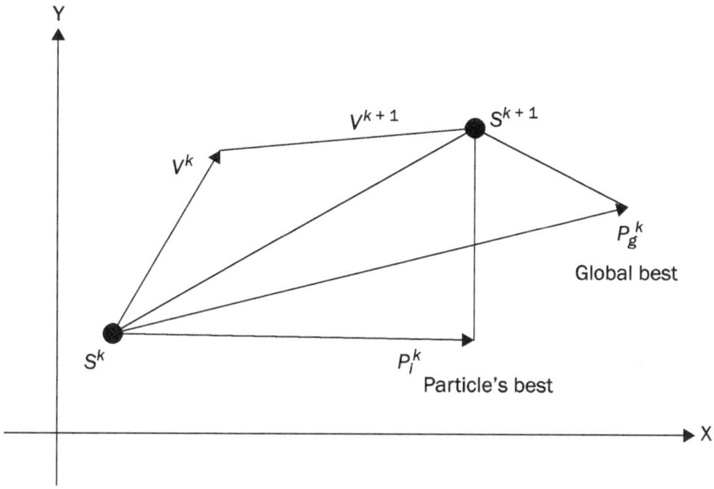

to maintain the diversity of the population and are uniformly distributed in the range [0–1]. The concept of modification of a searching point by PSO in a two-dimensional space is illustrated in Figure 7.1.

The PSO algorithm is based on a sociometric principle called g_{best}, which connects all the members of the swarm to one another (Kennedy and Eberhart, 2001). Each particle is affected by the very best performance of any member of the whole population. The key operators involved in evolutionary algorithm (EA) are recombination, mutation and selection; however, PSO does not have a direct recombination operator (Kennedy and Eberhart, 2001). But, the stochastic acceleration of a particle towards its previous best position as well as towards the best particle of the swarm resembles the recombination in EA. In PSO, the information exchange takes place merely among the particle's own experience and the experience of the best particle in the swarm instead of being carried from fitness-dependent selected parents to descendants as in genetic algorithms (GA). The directional position updating in PSO is analogous to mutation of GA. The PSO does not make use of the '*survival of the fittest*' concept and does not utilise a direct selection function. Hence, the particles with lower fitness can stay alive during the optimisation and potentially visit any point of the search space (Kennedy and Eberhart, 2001).

7.4 Experimental details

7.4.1 Planning of experiments

A careful planning of experimentation is necessary to develop the burr size mathematical models based on RSM. In the current research, cutting speed (v), feed rate (f), drill diameter (d) and point angle (θ) are identified as the process parameters, which affect the burr size, namely, burr height (B_h) and burr thickness (B_t). The process parameters and their levels were chosen based on preliminary experiments performed on drilling of stainless steel work material using HSS twist drills. The authors' preliminary study suggests that the cutting speed in the range of 8–16m/min is appropriate for controlling the burr formation. It was also observed during preliminary experimentation that the drilling performance decreased beyond feed rate of 0.12mm/rev due to excessive temperature rise and hence feed values were chosen in the range 0.04–0.12mm/rev. In industries, usually, the drilling of stainless steel work material is carried out with HSS drills with higher point angle for bigger diameter values. Further, the investigations carried out by Stein (1997) also recommended higher point angles for drilling stainless steel and hence the point angle in the range 118–134 degrees has been selected in the present investigation.

The drilling experiments were planned as per Box-Behnken design, which is an independent quadratic design that does not contain factorial or fractional factorial design and significantly reduces the number of experiments. The treatment combinations in Box-Behnken design are at the midpoints of edges of the process space and at the centre it requires three levels of each factor and is rotatable. The process parameters and the levels defined for each of the four parameters are given in Table 7.1. The Box-Behnken design of our current study consists of 27 sets of input process parameter combinations and is presented in Table 7.2.

Table 7.1 **Process parameters and their levels**

Parameter	Notation	Unit	Levels		
Cutting speed	*v*	m/min	8	12	16
Feed	*f*	mm/rev	0.04	0.08	0.12
Drill diameter	*d*	mm	16	22	28
Point angle	θ	degree	118	126	134

Table 7.2 **Box-Behnken design (BBD) matrix and measured values of burr size**

Trial No.	Process parameter settings				Measured values of burr size	
	Cutting speed (m/min)	Feed (mm/rev)	Drill diameter (mm)	Point angle (degree)	Burr height (mm)	Burr thickness (mm)
1	16	0.08	28	126	0.345	0.125
2	12	0.12	28	126	0.500	0.144
3	8	0.08	28	126	0.243	0.128
4	12	0.08	28	134	0.052	0.183
5	12	0.04	22	118	0.461	0.120

(*Continued*)

Table 7.2 *Continued*

Trial No.	Process parameter settings				Measured values of burr size	
	Cutting speed (m/min)	Feed (mm/rev)	Drill diameter (mm)	Point angle (degree)	Burr height (mm)	Burr thickness (mm)
6	8	0.04	22	126	0.094	0.160
7	12	0.08	22	126	0.174	0.180
8	16	0.08	16	126	0.164	0.186
9	12	0.12	22	134	0.367	0.217
10	8	0.08	16	126	0.130	0.152
11	12	0.08	22	126	0.174	0.180
12	16	0.12	22	126	0.367	0.187
13	12	0.12	22	118	0.562	0.172
14	12	0.08	16	118	0.203	0.141
15	12	0.04	22	134	0.061	0.247
16	8	0.08	22	134	0.049	0.228
17	16	0.08	22	134	0.103	0.201
18	12	0.08	16	134	0.214	0.197
19	12	0.04	28	126	0.364	0.144
20	8	0.12	22	126	0.440	0.202
21	8	0.08	22	118	0.331	0.099
22	16	0.04	22	126	0.305	0.206
23	16	0.08	22	118	0.415	0.158
24	12	0.04	16	126	0.150	0.175
25	12	0.08	22	126	0.174	0.180
26	12	0.08	28	118	0.658	0.069
27	12	0.12	16	126	0.421	0.198

7.4.2 Experimentation and burr size measurement

The drilling experiments were carried out as per Box-Behnken design on a three-axis '*YCM- V116B*' CNC vertical machining centre (Make: Yeong Chin Machinery Industries Co., Taiwan). The machining centre has a Fanuc controller with 15kW drive motor, feed rate of 5,000mm/min (maximum) and a spindle speed in the range 45–4,000rpm. The table travels along X-axis: 1,100mm (maximum); Z-axis: 630mm; and saddle travels along Y-axis: 600mm (maximum). The drilling fixture was mounted in the vice on the machine tool table and was used to clamp the work-pieces on a flat surface to maintain the perpendicularity of the exit surface of the hole and the machine spindle.

AISI 316L stainless steel work-pieces 25mm thick were used for the entire investigation; the chemical composition and mechanical properties are listed in Table 7.3. The parallel shank stub series HSS twist drills (Make: Addison & Co. Ltd., India) conforming to IS: 5100/DIN: 1897/BS: 328/ISO specifications were employed for the drilling experiments and the different point angles were ground as per the required experimental layout plan of Box-Behnken design for different sizes of the drills. The '*Cut60EP*' water-soluble oil was used as coolant throughout the investigation. The work specimens were polished on the exit surface before drilling for the preparation of burr size measurements.

Table 7.3 **Chemical composition and mechanical properties of AISI 316L stainless steel work material**

Chemical composition (wt %)	**0.026 C, 1.6 Mn, 0.37 Si, 0.029 P, 16.55 Cr, 10.0 Ni, 2.02 Mo, 0.16 Co, 0.027 S, 0.036 N**
Mechanical properties	Yield strength: 300–331MPa Tensile strength: 582–591MPa Hardness: 170BHN Elongation: 53% Reduction of area: 73%

The *'RPP-400'* toolmakers' microscope (Make: Sicherun-Gen Versehen, Germany) of 1μm resolution and 30× magnification was employed to measure the burr height and burr thickness. To measure the burr height, in the beginning, the focus was put on top of the burr and then on the exit surface and the distance between the two foci is the burr height. The burr thickness was measured by reading the distance between the inner surface of the drilled hole and the location at which the curvature of the root of the burr begins. The burr size values were recorded at four equally spaced locations around the circumference and the average reading was taken as the process response. The experimental layout plan as per Box-Behnken design and the measured values of burr height and burr thickness are summarised in Table 7.2.

7.5 Results and discussion

7.5.1 Development of mathematical models for burr size

The multiple regression analysis was performed for constructing the RSM-based burr size mathematical models. In the present investigation, the second order burr height and burr thickness mathematical models have been developed in terms of four process parameters, namely, cutting speed (v), feed rate (f), drill diameter (d) and point angle (θ). The non-linear second order response surface equation is given by:

$$Y = b_0 + b_1 v + b_2 f + b_3 d + b_4 \theta + b_{12} vf + b_{13} vd + b_{14} v\theta + b_{23} fd + b_{24} f\theta + b_{34} d\theta + b_{11} v^2 + b_{22} f^2 + b_{33} d^2 + b_{44} \theta^2 \quad (7.6)$$

where Y is the response, i.e. burr height (B_h) or burr thickness (B_t), and $b_o, \ldots, b_{44}$ are the regression coefficients, which are to be determined using Equation 7.3. The mathematical models as determined by regression analysis to predict the burr height (B_h) and burr thickness (B_t) are given by:

$$\begin{aligned} B_h = {} & 8.846555 + 0.066094v - 22.5026f + 0.356476d \\ & - 0.177890\theta - 0.44375vf + 0.000708vd - 0.00023v\theta \\ & - 0.14062fd + 0.160156f\theta - 0.00321d\theta - 0.00034v^2 \\ & + 83.04688f^2 + 0.001441d^2 + 0.000873\theta^2 \end{aligned} \tag{7.7}$$

$$\begin{aligned} B_t = {} & -2.9383 + 0.109156v + 8.563542f - 0.0008d \\ & + 0.028794\theta - 0.09531vf - 0.00039vd - 0.00067v\theta \\ & - 0.02396fd - 0.064066f\theta + 0.000302d\theta - 0.00027v^2 \\ & + 8.255208f^2 - 0.00078d^2 - 0.000067\theta^2 \end{aligned} \tag{7.8}$$

where v is in m/min; f is in mm/rev; d is in mm; θ is in degrees; B_h and B_t are in mm.

7.5.2 Adequacy checking of burr size models

The adequacy of the developed RSM-based burr size mathematical models has been checked through the analysis of variance (ANOVA) method (Montgomery, 2003). The ANOVA table consists of sum of squares, degrees of freedom and mean square. The sum of squares is performed into contributions from the regression model and the residual. The mean square is obtained by dividing the sum of squares of each of the sources of variation by the respective degrees of freedom. The Fisher's ratio (F-ratio) is the ratio of mean square of lack of fit to the mean square of pure residual error. As per ANOVA, the calculated value of F-ratio of developed model should be more than the Fisher's table (F-table) for the model to be adequate. Table 7.4 presents the summary of ANOVA and it is found that the burr size models are highly significant at 99% confidence interval.

Table 7.4 Analysis of variance (ANOVA) for burr size models

Burr size	Sum of squares		Degrees of freedom		Mean square		*F-ratio*
	Regression	Residual	Regression	Residual	Regression	Residual	
Burr height	0.714311	0.000001	14	12	0.051022	0	524800.29
Burr thickness	0.0416250	0.0000015	14	12	0.0029732	0.0000001	23785.72
$\text{F-Table}_{(14,\ 12,\ 0.01)} = 4.06$							

The goodness of fit of the burr size mathematical models was also tested through the coefficient of determination (R^2) and adjusted coefficient of determination (R^2_{adj}) (Montgomery, 2003). The R^2 is the proportion of variation in the response explained by the regression model. In contrast, R^2_{adj} is the coefficient of determination adjusted for the number of variables in the regression model. Unlike R^2, the R^2_{adj} may decrease if the variables are entered in the model that does not add drastically to the model fit. The coefficient of determination is given by (Montgomery, 2003):

$$R^2 = \frac{SS_R}{SS_{yy}} \tag{7.9}$$

where the sum of squares due to regression is

$$SS_R = \sum_i^n (y_{i,pred} - \bar{y}_{i,\mathrm{expt}})^2 \tag{7.10}$$

where $y_{i,expt}$ is the measured response corresponding to data set i; $y_{i,pred}$ is the predicted response corresponding to data set i; $\bar{y}_{i,expt}$ is the overall average of $y_{i,expt}$ and n is the number of data sets in the Box-Behnken design. The sum of squares about the mean is given by:

$$SS_{yy} = \sum_i^n (y_{i,expt})^2 - \frac{\left(\sum_i^n y_{i,expt}\right)^2}{n} \tag{7.11}$$

The adjusted coefficient of determination is given by (Montgomery, 2003):

$$R_{adj}^2 = 1 - \frac{SS_R / (n-p)}{SS_{yy} / (n-1)} \tag{7.12}$$

where p is the total number of regressors (excluding the constant term) in the regression model. If R^2 value is closer to 1, then the model is strong and it predicts the response better (Montgomery, 2003). The R^2 and R^2_{adj} values for both the burr height and burr thickness models were found to be 1, which clearly indicates the strong correlation between the experimental and the predicted values of the responses.

7.5.3 Accuracy determination of burr size models

Equations 7.7 and 7.8 are used to test the accuracy of the developed mathematical models using the experimental data of Box-Behnken design as well as the validation data set. The percentage prediction accuracy is given by:

$$\delta = \frac{100}{n} \times \sum_{i=1}^{n} \left| \frac{y_{i,expt} - y_{i,pred}}{y_{i,pred}} \right| \tag{7.13}$$

where $y_{i,expt}$ is measured value of response corresponding to i^{th} trial; $y_{i,pred}$ is predicted value of response corresponding to i^{th} trial and n is the number of trials. The average prediction error rate of the developed RSM-based mathematical models with 27 sets of experimental data of Box-Behnken design is found to be 7.3 per cent and 0.91 per cent for the burr height and burr thickness respectively. The comparisons of the experimental and the predicted values of burr height and burr thickness for the experimental data of Box-Behnken design are shown in Figures 7.2 and 7.3 respectively.

In order to validate the developed burr size models, five sets of random experiments were performed within the ranges of the process parameters selected and the burr height and burr thickness values were measured. The experimental values were then compared with the predicted values of RSM-based models and it was found that these values were close to each other, as illustrated in Table 7.5. The maximum prediction

Figure 7.2 Experimental and predicted values of burr height for the experimental data

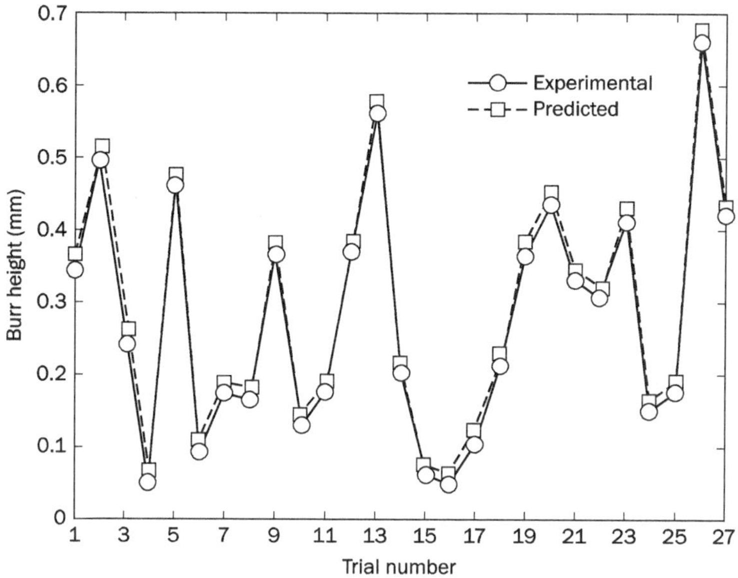

Figure 7.3 Experimental and predicted values of burr thickness for the experimental data

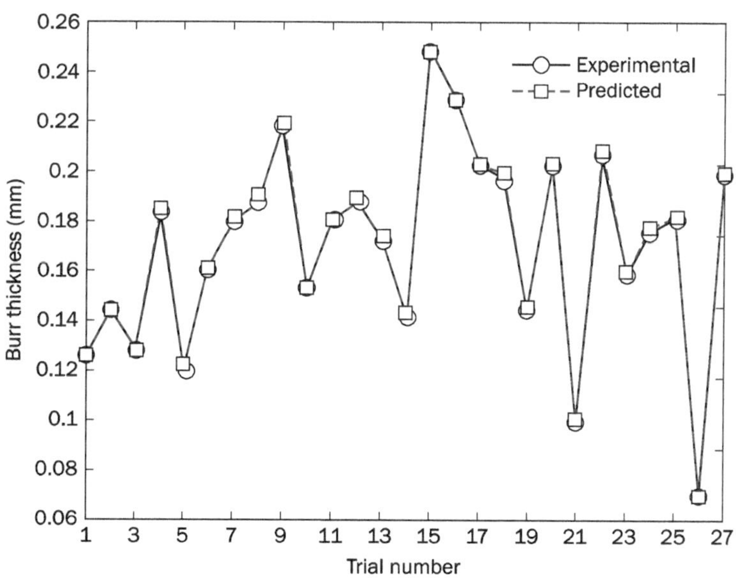

Table 7.5 Comparison of experimental and RSM-based predicted values of burr size for the validation data

S. No.	Process parameter settings				Experimental values		Predicted values		Prediction error	
	v (m/min)	f (mm/rev)	d (mm)	θ (degree)	B_h (mm)	B_t (mm)	B_h (mm)	B_t (mm)	B_h (%)	B_t (%)
1	16	0.08	20	120	0.331	0.182	0.304	0.178	8.88	2.25
2	12	0.05	18	130	0.113	0.192	0.116	0.207	2.59	7.25
3	10	0.06	22	132	0.042	0.231	0.039	0.219	7.69	5.48
4	8	0.07	16	118	0.141	0.102	0.152	0.088	7.24	15.91
5	14	0.04	26	134	0.131	0.254	0.125	0.232	4.80	9.48

errors of 8.88% and 15.91% were observed for burr height and burr thickness respectively for the validation data set.

7.5.4 Parametric analysis of process parameters on burr size

The main purpose of the current study is to minimise the burr size for a specified drill diameter and hence the influence of combination of drill diameter along with the remaining process parameters such as cutting speed, feed and point angle is of concern. Therefore, 3D surface plots (Figures 7.4, 7.5 and 7.6) were generated using second-order RSM-based mathematical models to analyse the two-factor interaction effects of process parameters, namely, drill diameter-cutting speed, drill diameter-feed and drill diameter-point angle on burr size. The surface plots were constructed using MATLAB (Math Works Incorporation, 2005) considering two parameters at a time, while the third parameter was kept at the centre level. The response surface analysis of two-factor interaction effects conveys important information on burr behaviour with respect to drilling factors.

The interaction effect due to drill diameter and cutting speed on burr size is exhibited in Figure 7.4. It is observed that the burr height is sensitive to drill diameter variations for a specified cutting speed (Figure 7.4a). Moreover, it is highly sensitive at higher cutting speed as compared with lower cutting speed values. Therefore, for all the drill diameters specified, lower cutting speeds will result in smaller burr heights. On the other hand, the interaction effect of drill diameter and cutting speed shows a different behaviour for burr thickness (Figure 7.4b). It is evident from this figure that the variation of burr thickness with respect to drill diameter is almost the same for all the cutting speeds.

Figure 7.4 Interaction effect of drill diameter and cutting speed on burr size (f = 0.08 mm/rev and θ = 126 degrees)

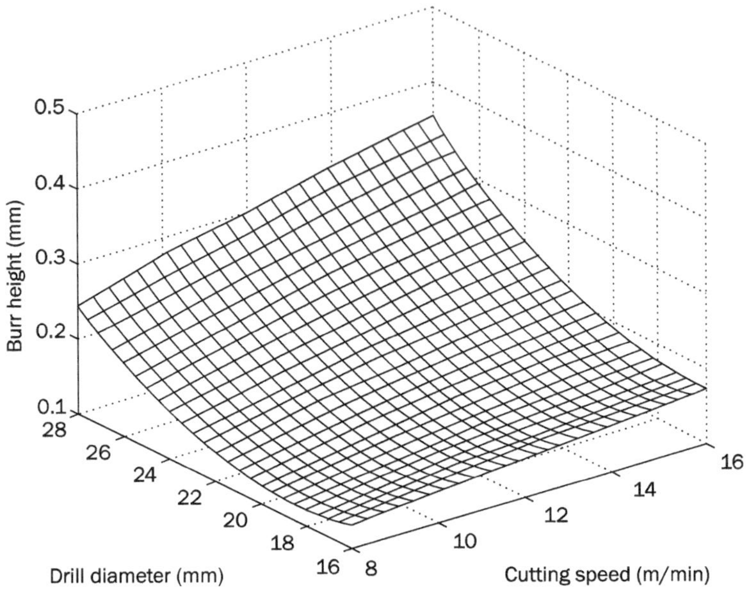

(a)

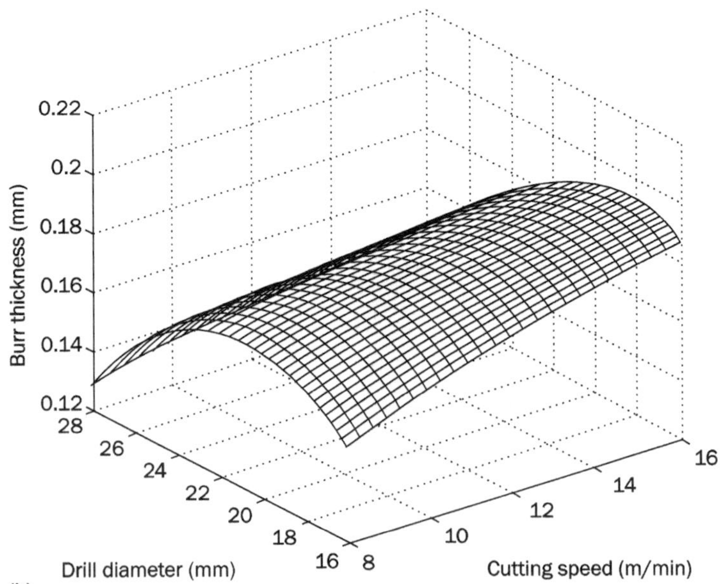

(b)

Figure 7.5 Interaction effect of drill diameter and feed on burr size (v = 12m/min and θ = 126 degrees)

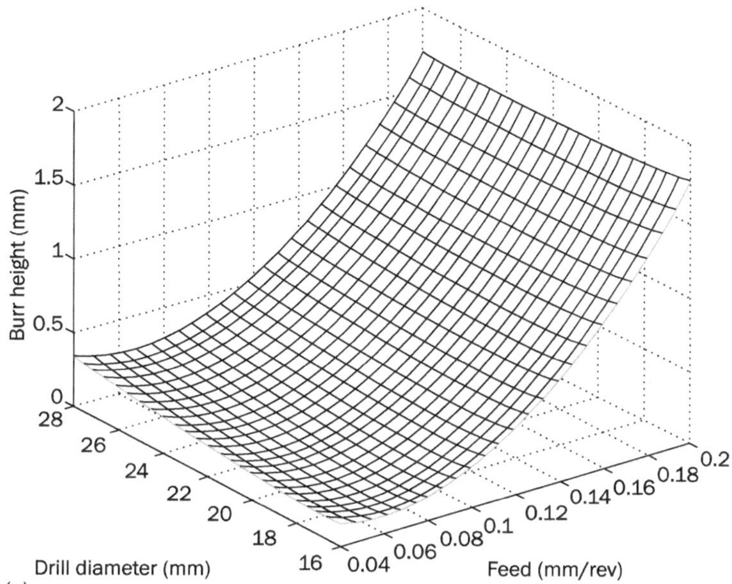

(a)

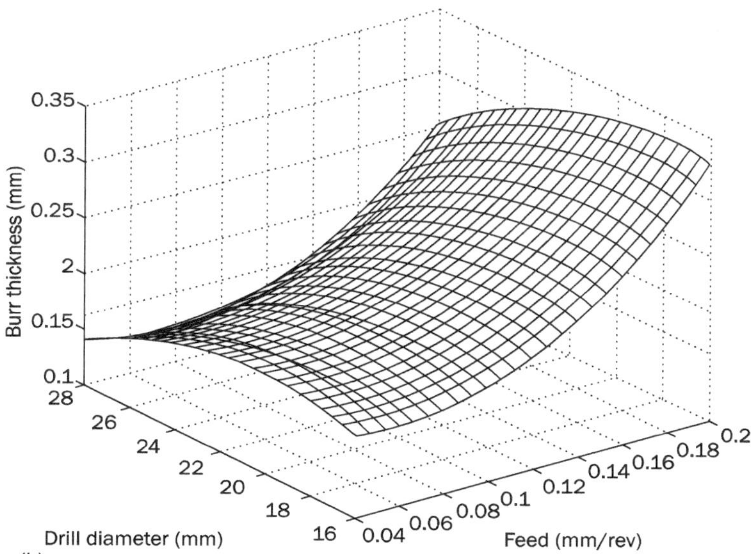

(b)

Figure 7.6 **Interaction effect of drill diameter and point angle on burr size (v = 12m/min and f = 0.08mm/rev)**

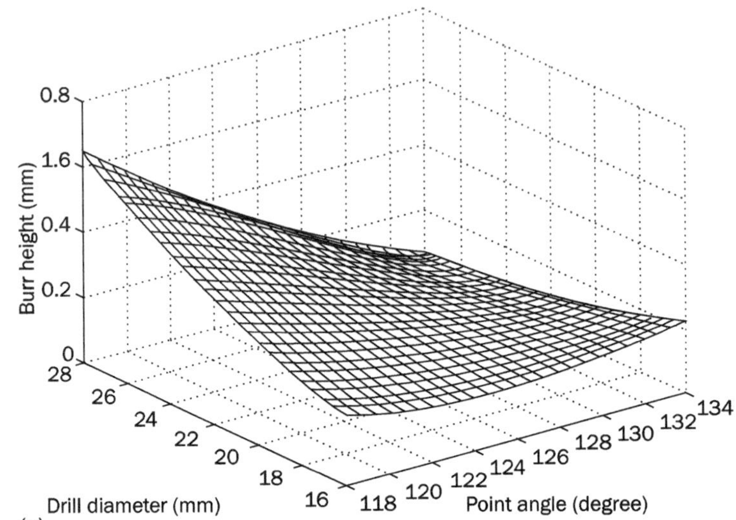

(a)

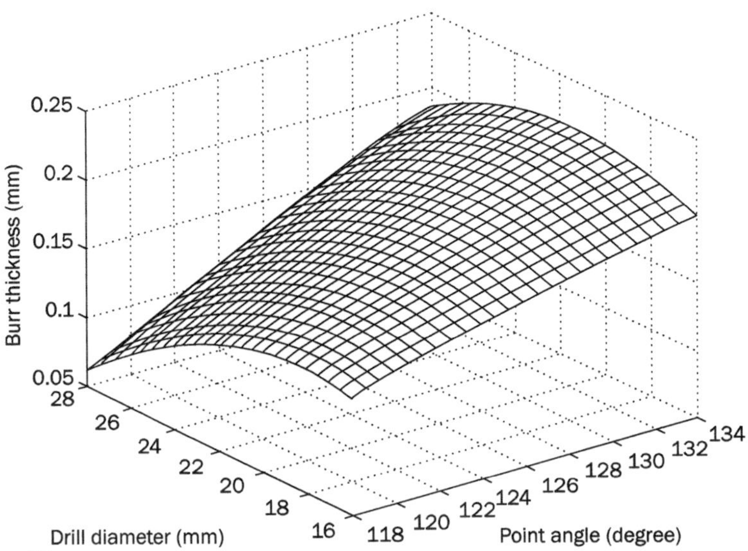

(b)

However, lower cutting speed results in smaller burr thickness for all the drill diameters specified in the range 16–28mm.

As seen from Figures 7.5a and 7.5b, the interaction effect of feed and drill diameter is almost the same for both the burr height and burr thickness. For a given drill diameter, increased feed results in a sharp increase in both burr height and burr thickness. This indicates that the burr size is highly sensitive to the variations in feed and it is necessary to employ lower feed to minimise the burr size.

The variation of burr size with respect to drill diameter and point angle is displayed in Figure 7.6. The burr height is highly sensitive to point angle variations for drilling the holes of larger diameter. It is clear from this interaction effect (Figure 7.6a) that, as the drill diameter is increased, it is essential to simultaneously increase the point angle so as to minimise the burr height. However, this has an opposing effect on burr thickness, which increases with the simultaneous increase of drill diameter and point angle (Figure 7.6b). Irrespective of the drill diameter, lower point angle results in smaller burr thickness.

As can be seen from the above surface plots, there exists a non-linear relationship between the burr size and the process parameters. However, the behaviour of the burr height is entirely different from that of the burr thickness. For a given drill diameter, a combination of cutting speed, feed and point angle, which minimises the burr height, may not be appropriate for minimising the burr thickness. Hence, different combinations of cutting speed, feed and point angle are required to simultaneously minimise the burr height and burr thickness at different drill diameter values. Thus PSO, an efficient multi-objective optimisation tool, has been selected for determining the optimal combination of cutting speed, feed and point angle for a specified drill diameter in the range 16–28mm for simultaneously minimising the burr height and burr thickness.

7.5.5 Development of PSO for burr size minimisation

The RSM-based burr height and burr thickness models were employed to determine the best combination of process parameters, namely, cutting speed (v), feed (f) and point angle (θ) for a specified drill diameter (d), which results in minimal burr height and burr thickness. In the present investigation, the fitness function is designed by linear mapping of the predicted burr height (B_{hp}) and burr thickness (B_{tp}) and is given by:

$$fit = 100 - (50B_{hp} + 100B_{tp}) \tag{7.14}$$

This linearly maps burr height values from 0 to 1mm and burr thickness from 0 to 0.5mm into fitness range [100:0] as represented in Figure 7.7. Here, B_{hp} and B_{tp} are the burr height and burr thickness values predicted by RSM-based models for a specified drill diameter (d) using Equations 7.7 and 7.8 respectively. In our investigation, the simultaneous minimisation of B_h and B_t requires the maximisation of fit and hence the constrained optimisation problem is:

- Determine the optimal values of v, f and θ for a specified d.
- So as to maximise fit.
- Subject to the constraints: $8 \leq v \leq 16$ m/min; $0.04 \leq f \leq 0.12$ mm/rev; $118 \leq \theta \leq 134$ degree.

The various steps involved in PSO of our investigation are, for a specified drill diameter (d) in the range 16–28mm:

- Step 1: Randomly initialise m sets of parameters (particles), namely, cutting speed (v), feed (f) and point angle (θ). The

Figure 7.7 Fitness mapping of burr height and burr thickness

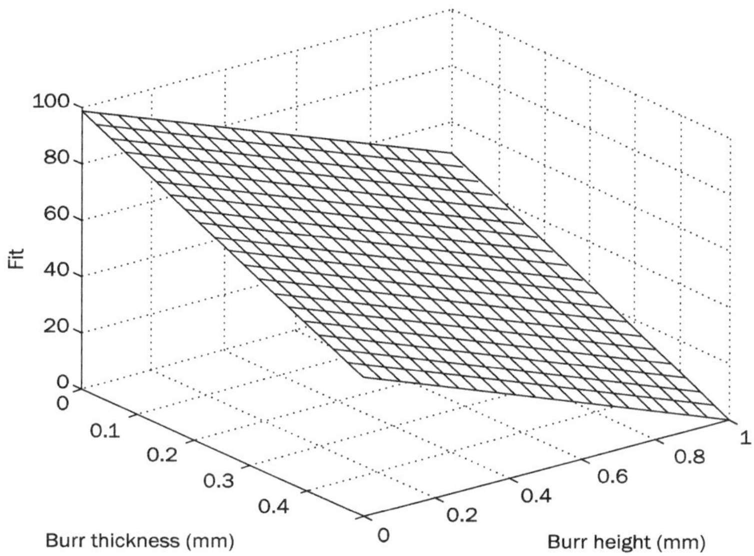

current particles of every set are positioned to 'p_{best}' and the parameters corresponding to the best fitness among all the sets are chosen as 'g_{best}'.

- Step 2: With specified value of drill diameter (d), determine the predicted values of B_{hp} and B_{tp} using the RSM-based models of Equations 7.7 and 7.8 respectively.
- Step 3: Compute the fitness of every set of process parameters using Equation 7.14.
- Step 4: Compare the calculated fitness with the fitness corresponding to 'p_{best}', if the fitness is greater than 'p_{best}', then 'p_{best}' should be updated.
- Step 5: Compare the value of maximum fitness with the 'g_{best}' and if the maximum fitness is greater than 'g_{best}', then 'g_{best}' should be updated.
- Step 6: Update the position and velocity of the particles using Equations 7.4 and 7.5.

- Step 7: Judge whether the programme will stop (total iteration is usually set as a termination rule). If true, stop the iteration, otherwise go back to Step 3.

7.5.6 PSO optimisation results

The PSO simulation was performed using MATLAB software (Math Works Incorporation, 2005) with the maximum number of 100 generations (k_{max}). For every value of drill diameter (d) specified in the range 16–28mm, the best combination of cutting speed (v), feed (f) and point angle (θ) and the corresponding optimal burr height (B_h) and burr thickness (B_t) were recorded. In PSO, the typical range for the number of particles is preferred between 20 and 50 and hence in the current study, the swarm size is selected as 50. The values of the learning factors c_1 and c_2 are taken as 2.0. Initially, a large value of inertia weight (w_{max}) was taken to allow for a global search. In order to reduce this weight over the iterations allowing the algorithm to exploit some specific areas, the following equation is used:

$$w^k = w_{\max} - \left[\frac{(w_{\max} - w_{\min})k}{k_{\max}} \right] \tag{7.15}$$

Here w_{max} = 0.80 and w_{min} = 0.01. The PSO convergence characteristic for a drill diameter of 20mm is depicted in Figure 7.8.

The optimal values of burr height and burr thickness for various drill diameters in the range 16–28mm are displayed in Figure 7.9. As seen from this figure, the optimal burr height decreases with the increase in drill diameter up to 24mm and then remains almost constant beyond 24mm drill diameter. On the other hand, the optimal burr thickness increases with the increase in drill diameter up to 24mm and then starts decreasing beyond 24mm.

Figure 7.8 PSO convergence for a drill diameter of 20mm

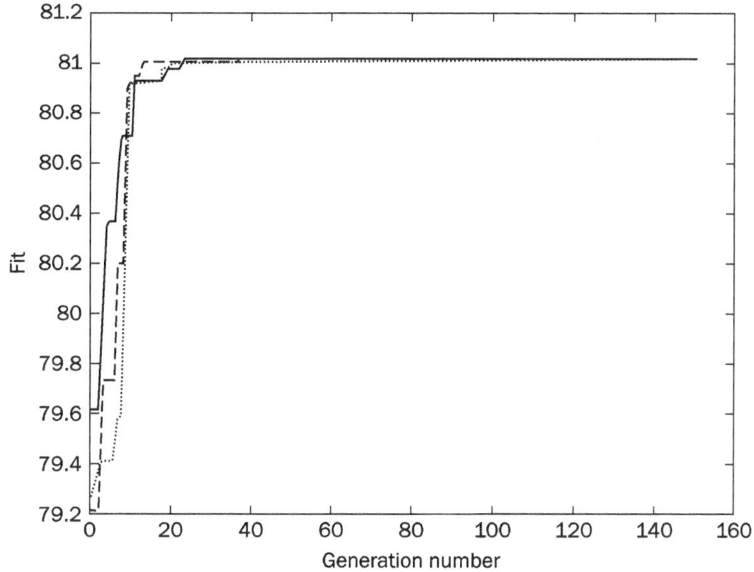

Figure 7.9 Optimal burr size obtained by PSO simulation for different values of drill diameter

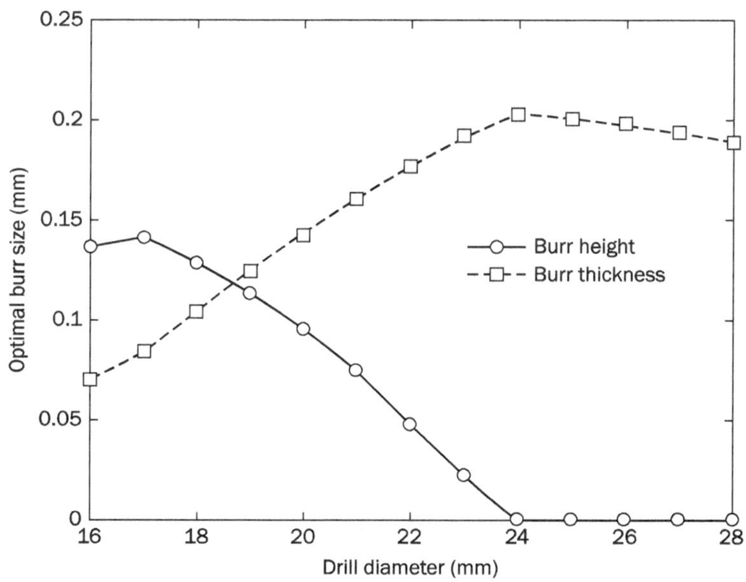

It is observed from the optimisation results that for minimising burr size, the optimal cutting speed requirement is low, i.e. 8m/min irrespective of the drill diameter. On the other hand, as depicted in Figure 7.10, the requirement of optimal feed is in the low range (0.04–0.06mm/rev) and is in direct proportion to the drill diameter for the entire range specified. The reason might be that at lower feed thrust force reduces and hence forms thinner burrs.

The relationship between the drill diameter and the optimal point angle is shown in Figure 7.11. As can be seen from this figure, the optimal point angle requirement is 118 degrees for smaller drill diameter of 16mm and 132 degrees for larger drill diameter of 28mm for minimising the burr size. However, the optimal point angle is directly proportional to the drill diameter and it is always advantageous to employ larger point angles for bigger drill diameter values. This is

Figure 7.10 Relationship between drill diameter and optimal feed

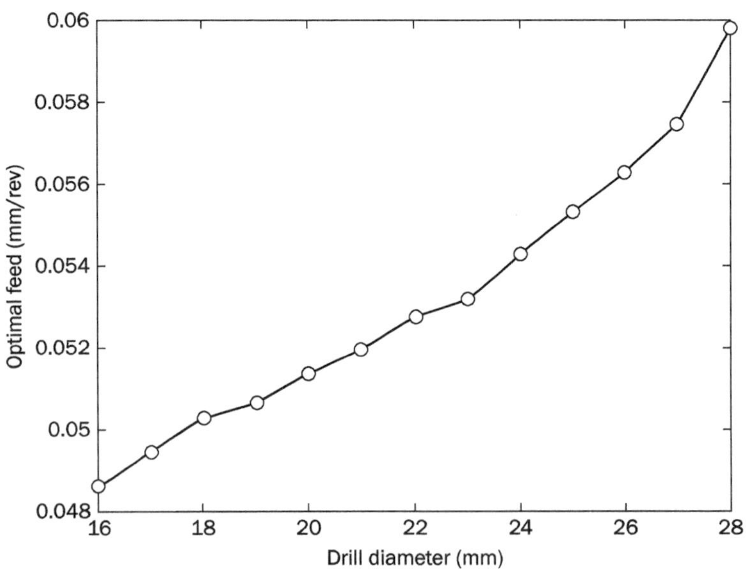

Figure 7.11 **Relationship between drill diameter and optimal point angle**

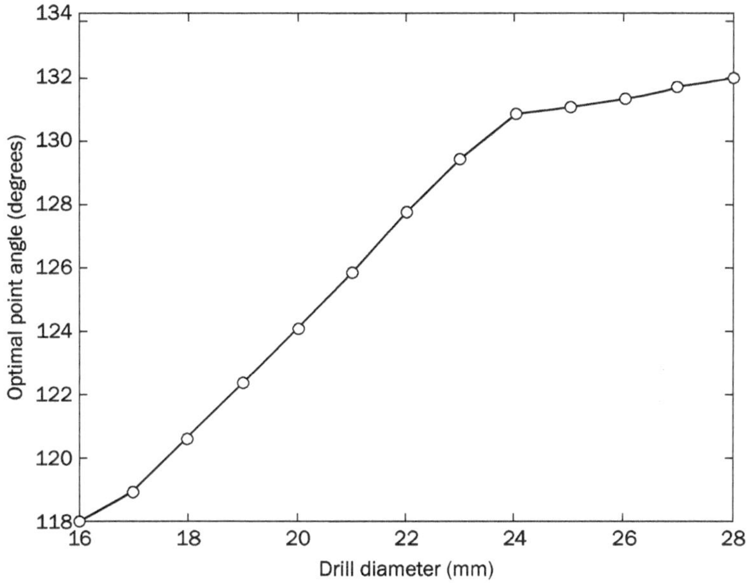

due to the fact that higher point angle ensures maximum lip movement in the earliest possible time to avoid strain hardening and thus due to change in the chip flow direction the thinner burrs will be produced with an increased point angle. Further, with the bigger diameter values and the corresponding increased point angle, the drill generates a disk at the hole end and keeps the work under tensile stress that causes the material to cut easily and hence forms thinner burrs.

From the optimisation results, it is observed that the combination of optimal feed and point angle has played a key role in reducing the burr size for every drill diameter specified in the range 16–28mm. However, the different combinations of optimal feed and point angle are required for minimising the burr size. It is also revealed that, with the increase in drill diameter, simultaneous increase in feed and point angle is necessary for controlling the burr size, whereas the optimal cutting speed requirement is 8m/min irrespective of the drill diameter. The optimisation results demonstrated in the current study are one such possible optimal solution and there are multiple optimal solutions, as the relationship between the burr size and the process parameters is highly non-linear. The parameters selected for our investigation are chosen after several trials of PSO simulation and are found to be appropriate for the drilling optimisation.

7.6 Conclusions

The application of particle swarm optimisation (PSO) for simultaneous minimisation of burr height and burr thickness during drilling of AISI 316L stainless steel has been presented in this chapter. The machining tests were performed with a plan of experiments based on Box-Behnken quadratic design. The burr size second-order mathematical models were developed in terms of cutting speed, feed, drill diameter and point angle using response surface methodology (RSM). The developed models were then tested through the analysis of variance (ANOVA) and found to be adequate at 99 per cent confidence interval. The fitness function was designed by linear mapping of burr height and burr thickness. For every value of drill diameter specified, the best combination of the process parameters has been determined through PSO simulation. The following conclusions are drawn from the present investigation:

- The existence of non-linear relationship between the burr size and the identified process parameters is clearly evidenced in the response surface plots and hence justifying the use of RSM.

- The burr height is highly sensitive at higher values of cutting speed as compared with lower cutting speeds. For all the drill diameters specified, lower cutting speeds result in thinner burrs.
- The burr size is highly sensitive to feed variations and it is beneficial to employ lower feed to minimise the burr size.
- The burr height decreases with the increase in point angle for bigger drill diameter values. However, the burr thickness increases with the increase in drill diameter.
- The different combinations of feed and point angle are required to minimise the burr size for different drill diameter values specified in the range 16–28mm.
- The requirement of optimal cutting speed is 8m/min for reducing the burr size irrespective of the drill diameter.
- The optimal feed requirement is low (0.04–0.06mm/rev) and is directly proportional to the drill diameter in order to minimise the burr size.
- The optimal point angle requirement is also in direct proportion to the drill diameter and it is useful to employ larger point angles for bigger drill diameter values.
- With the increase in drill diameter, simultaneous increase in feed and point angle is desirable for controlling the burr size.

7.7 References

Arai, M. and Nakayama, K. (1986) Boundary notch on cutting tool caused by burr and its suppression, *Bulletin of the Japan Society of Precision Engineering*, 52: 864–6.

Dornfeld, D.A. (1992) Intelligent deburring of precision components, *Proceedings of the International Conference on Industrial Electronics, Control, Instrumentation and Automation, IEEE*, San Diego, Vol. 2: 953–60.

Dornfeld, D.A., Kim, J., Dechow, H., Hewson, J. and Chen, I.J. (1999) Drilling burr formation in titanium alloy, Ti-6Al-4V, *Annals of CIRP*, 52(1): 45–8.

Elbeltagi, E., Hegazy, T. and Grierson, D. (2005) Comparison among five evolutionary-based optimization algorithms, *Advanced Engineering Information*, 19(1): 43–53.

Gaitonde, V.N., Karnik, S.R., Achyutha, B.T. and Siddeswarappa, B. (2007) Methodology of Taguchi optimization for multi-objective drilling problem to minimize burr size, *International Journal of Advanced Manufacturing Technology*, 34: 1–8.

Gaitonde, V.N. and Karnik, S.R. (2007) Taguchi robust design for multi-response drilling optimization to minimize burr size using utility concept, *International Journal of Manufacturing Research*, 2(2): 209–24.

Gaitonde, V.N., Karnik, S.R., Achyutha, B.T. and Siddeswarappa, B. (2008) Taguchi optimization in drilling of AISI 316L stainless steel to minimize burr size using multi-performance objective based on membership function, *Journal of Materials Processing Technology*, 202: 374–9.

Gaitonde, V.N., Karnik, S.R., Achyutha, B.T., Siddeswarappa, B. and Davim, J.P. (2009) Predicting burr size in drilling of AISI 316L stainless steel using response surface analysis, *International Journal of Materials and Product Technology*, 35(1–2): 228–45.

Gillespie, L.K. (1996) *Standard Terminology for Researchers of Burrs and Edge Finishing*; WBTC STD-02.

Guo, Y.B. and Dornfeld, D.A. (2000) Finite element modeling of drilling burr formation process in drilling 304 stainless steel, *ASME Journal of Manufacturing Science and Engineering*, 122(4): 612–19.

Hunag, M.F. (2004) Application of grey-Taguchi method to optimize drilling of aluminium alloy 6061 with multiple performance characteristics, *Materials Science and Technology*, 20: 528–32.

Kennedy, J. and Eberhart, R.C. (2001) *Swarm Intelligence*. San Diego, CA: Academic Press.

Kim, J. and Dornfeld, D.A. (2000) Development of a drilling burr control chart for low alloy steel, AISI 4118, *Journal of Materials Processing Technology*, 113: 4–9.

Kim, J., Min, S. and Dornfeld, D.A. (2001) Optimization and control of drilling burr formation of AISI 304L and AISI 4118 based on drilling burr control charts, *International Journal of Machine Tools and Manufacture*, 41: 923–36.

Ko, S.L. and Lee, J. (2001) Analysis on burr formation in drilling with new concept drill, *Journal of Materials Processing Technology*, 113: 392–8.

Koelsch, J. (2001) Divining edge quality by reading the burrs, *Quality Magazine*, 24–8.

Konstantinos, E., Parsopoulos, M.N. and Vrahatis (2004) On the computation of all global minimizers through particle swarm optimization, *IEEE Transactions on Evolutionary Computing*, 8(3): 211–24.

Lauderbaugh, L.K. (2009) Analysis of the effects of process parameters on exit burrs in drilling using a combined simulation and experimental approach, *Journal of Materials Processing Technology*, 209: 1909–19.

Lin, R. (2002) Cutting behavior of a TiN-coated carbide drill with curved cutting edges during the high speed machining of stainless steel, *Journal of Materials Processing Technology*, 27: 8–16.

Math Works Incorporation (2005) *MATLAB User Manual*, Version 7.1, R 14, Natick, MA.

Min, S., Dornfeld, D.A., Kim, J. and Shyu, B. (2001a) Finite element modeling of burr formation in metal cutting, *Machining Science and Technology*, 5(2): 307–22.

Min, S., Kim, J. and Dornfeld, D.A. (2001b) Development of a drilling burr control chart for stainless steel, *Transactions of NAMRI/SME*, 28: 317–22.

Montgomery, D.C. (2003) *Design and Analysis of Experiments*. New York: John Wiley and Sons.

Saunders, L.K.L. (2003) A finite element modeling of exit burrs for drilling of metal, *Finite Element in Analysis and Design*, 40(2): 139–58.

Stein, J.M. (1997) *The Burrs from Drilling: An Introduction to Drilling Burr Technology*. Burr Technology Information Series ™: USA.

Stein, J.M. and Dornfeld, D.A. (1997) Burr formation in drilling miniature holes, *Annals of CIRP*, 46(1): 63–6.

Tosun, N. (2006) Determination of optimum parameters for multi-performance characteristics in drilling by using grey relational analysis, *International Journal of Advanced Manufacturing Technology*, 28: 450–5.

8

Single-point incremental forming of polymers

Maria Beatriz Silva, Tânia Marques and Paolo A.F. Martins

Abstract: This chapter focuses on new developments in single-point incremental forming (SPIF) that offer the possibility of producing low-cost, small-batch, polymer sheet components at room temperature. The presentation is supported by experimentation with benchmark shapes made from polyvinylchloride (PVC), polyamide (PA), polyethylene terephthalate (PET) and polycarbonate (PC) and is aimed at identifying the formability limits of the process as a function of the major operating parameters. These are further examined by means of a theoretical framework based on membrane analysis with in-plane contact friction forces that is capable of modelling the cold plastic deformation of polymers with pressure-sensitive yield surfaces. Results show that single-point incremental forming of commercial polymer sheets seems promising for the manufacture of complex sheet components with very high depths at room temperature.

Key words: single-point incremental forming; polymers; theoretical framework; experimentation.

8.1 Introduction

Polymers represent a significant percentage of the raw materials utilised in industry but their current processing techniques, running on either melting or rubbery states, are only suitable for mass production because the energy costs and the capital investment in equipment and tooling are very high. As a result of this, polymers are considered not to be suitable for flexible small-batch production (that is, low volume of custom and specific products) and, therefore, not appropriate for the growing agile manufacturing trends requiring very short life-cycles and very short development and production lead times.

The present need for flexible polymer processing techniques requires the development of innovative technologies that are capable of reducing the fixed and capital costs of polymer processing to a level where small-batch production becomes economically feasible. Although this conclusion may appear obvious it represents a serious challenge for the current state of the art of polymer processing due to the necessity of developing new technologies that are capable of operating at room temperature with low tooling costs.

In a recent paper, Franzen et al. (2008a) presented the first breakthrough towards the development of an innovative cold polymer processing technique for flexible small-batch production and rapid prototyping through the evaluation of the performance of commercial polyvinylchloride (PVC) sheets in single-point incremental forming (SPIF) applications (Figure 8.1). Several modes of deformation were identified and special emphasis was placed on the experimentally observed phenomena that are different from those commonly seen in the SPIF of metals.

In a subsequent paper Le et al. (2008) published a preliminary set of experimental results for the SPIF of

Figure 8.1 Single-point incremental forming: (a) schematic representation of the process; (b) tool set-up that was utilised in the experiments with polymer sheets

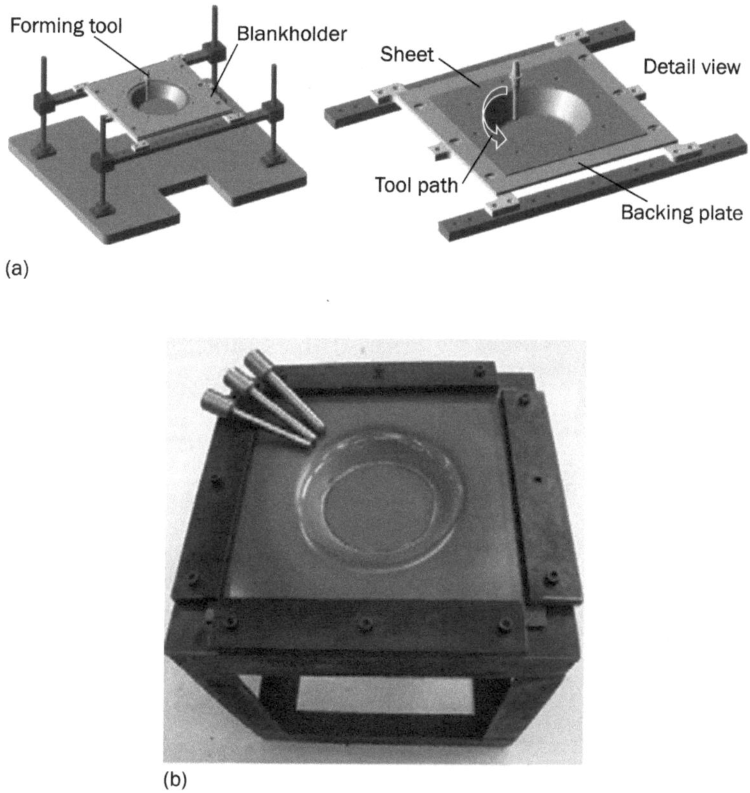

polypropylene (PP) sheets. The research work plan was built upon statistical analysis based on design of experiments and comprised the study of the influence of the step size, tool radius, feed rate and spindle speed on the overall formability of the process. Two failure modes similar to those previously observed by Franzen et al. (2008a) in PVC were reported.

In a second paper, Franzen et al. (2008b) extended the scope of their investigation to include four additional polymers: polyoxymethylene (POM), polyethylene (PE), polyamide (PA) and polycarbonate (PC), with material structures varying from high-crystalline (POM) to amorphous (PC). The previously identified failure modes were revisited and systematised in the light of new experimental data. A first attempt was made towards the development of criteria for the selection of polymers for SPIF applications.

More recently, Silva et al. (2010) provided a new and fundamental level of understanding about the SPIF of polymers by setting up a theoretical framework that allows the influence of major operative parameters and their mutual interaction to be studied both qualitatively and quantitatively. The theoretical framework is based on membrane analysis with in-plane contact friction forces and is capable of modelling the cold plastic deformation of polymers with pressure-sensitive yield surfaces.

This chapter is designed under the above-mentioned contributions to knowledge and presents the first steps towards the objective of extending the scope of application of SPIF to polymers. Table 8.1 presents a brief overview of the main technical and economic aspects of the conventional polymer processing technologies and forecasts the potential niche market for SPIF applications. In broader terms, this chapter is expected to provide answers to the following key questions related to SPIF of polymers that are crucial to predict its behaviour across the useful range of operative conditions and to exploit its potential for the manufacture of complex sheet parts with very high depths: What is the distribution of strain and stress in the small plastic zone created by the forming tool? What is the physics behind the different types of failure modes? How can the formability limits be experimentally determined? Is it possible to develop a unified theoretical framework for SPIF that includes both metals and polymers?

The presentation combines experimentation with conical and pyramidal benchmark shapes made from

Table 8.1 Technical and economic aspects of polymer processing

	Equipment cost	Tooling cost	Production rate	Batch size (number of parts)	Applications
Blow moulding	Medium	Medium	Medium	$>10^3$	Hollow thin-walled parts such as bottles, containers and fuel tanks
Compression moulding	High	Medium	Medium	$>10^3$	Parts similar to closed-die forgings primarily made from highly filled polymers with a high viscosity
Injection moulding	High	High	High	$>10^4$	Complex shapes of various sizes with fine details such as gears, fasteners, housings, switches and knobs
Transfer moulding	High	High	Medium	$>10^3$	More complex parts than those produced by compression moulding
Thermoforming	Low	Low	Low	$>10^2$	Parts with shallow or deep cavities made from polymer sheets such as packaging cases
SPIF	Medium	Low	Low	>1	Rapid prototypes and parts similar to those produced by thermoforming and made from polymer sheets

polyvinylchloride (PVC), polyamide (PA), polyethylene terephthalate (PET) and polycarbonate (PC) sheets with explanations of the results and observations provided by the aforementioned theoretical framework for SPIF of polymers.

8.2 Theoretical framework

The governing mode of deformation in SPIF of metals and the likely mode of failure have been subjects of considerable debate and controversy in academia. Some authors claim that deformation takes place by stretching instead of shearing while others claim the opposite (Emmens and van den Boogaard, 2007). In what concerns the formability limits, some authors explain the high formability of SPIF, and the corresponding raise of the forming limit curve (FLC), as a consequence of through thickness shear (Allwood et al., 2007; Jackson and Allwood, 2008) or serrated strain paths arising from cyclic, local plastic deformation (Eyckens et al., 2007). Others state that formability is limited by fracture due to the inability of necks to grow and that forming limit curves (FLCs), which give the loci of necking strains, are not relevant and should be replaced by the fracture forming limits (FFLs) (Silva et al., 2008b). The state-of-the-art review paper by Emmens and van den Boogaard (2009) presents an excellent overview of the most significant contributions in the field with special emphasis on the mechanisms that have been proposed to explain plastic deformation of metal sheets above the FLC.

However, as was recently shown by Franzen et al. (2008a, 2008b) and Silva et al. (2010), SPIF of polymers adds new phenomena in the characterisation of the deformation mechanics and formability limits by introducing new modes of failure and damage that are not experimentally observed in metals. This section starts with a brief overview of the fundamentals of the cold forming of polymers and follows with the presentation of a theoretical framework for the SPIF of polymers.

8.2.1 Cold forming of polymers

The fundamentals of cold forming of polymers have been comprehensively investigated since the mid-1960s. Whitney and Andrews (1967) proposed a pressure-modified Tresca yield criterion and Sternstein and Ongchin (1969) developed a first version of the pressure-modified von Mises yield criterion which was later enhanced by Raghava et al. (1973) and Caddell et al. (1974) in order to explicitly account for the differences between compressive σ_{YC} and tensile σ_{YT} flow stresses of a polymeric material (commonly known as the strength-differential (SD) effect) at a given strain. The yield function due to Raghava et al. (1973) and Caddell et al. (1974) is expressed as:

$$F\left(\sigma_{ij}\right)=\bar{\sigma}^2-\sigma_{YC}\cdot\sigma_{YT}+(\sigma_{YC}-\sigma_{YT})\sigma_{kk}=0 \qquad (8.1)$$

where $\bar{\sigma}=\sqrt{(3/2)\sigma'_{ij}\,\sigma'_{ij}}$ is the effective (or equivalent) stress and $\sigma_{kk}=\delta_{ij}\sigma_{ij}$. In case $\sigma_{YT}=\sigma_{YC}$, the traditional von Mises criterion (currently utilised in metals) results.

In what concerns incompressibility, several researchers stated after measuring volume variation of different polymers, that there is negligible volume variation after yielding and concluded that the normality rule, typical of associated plasticity, does not hold in case of the pressure dependent yield surfaces of polymers (Whitney and Andrews, 1967). This conclusion was confirmed by Spitzig and Richmond (1979) who showed that the associated flow rule based on a pressure sensitive yield surface leads to predictions of volume variation at least one order of magnitude larger than observed. As a result of this, the incompressibility of polymers in the plastic deformation regime can be taken into account by setting up constitutive equations built upon the following non-associated flow rule (Alves and Martins, 2009):

$$d\varepsilon_{ij} = d\lambda \frac{\partial Q(\sigma_{ij})}{\partial \sigma_{ij}} \tag{8.2}$$

where the plastic multiplier $d\lambda$ is given by (Lee, 1988):

$$d\lambda = \frac{3}{2}\frac{d\bar{\varepsilon}}{\bar{\sigma}} \tag{8.3}$$

and $d\bar{\varepsilon} = \sqrt{\tfrac{2}{3} d\varepsilon'_{ij}\, d\varepsilon'_{ij}}$ denotes the increment of the effective plastic strain.

By following an approach similar to that of the previous paragraphs, which starts by writing the effective stress in Equation 8.1 as $\bar{\sigma} = \sigma_1 - \sigma_3$, it is possible to set up the pressure-modified version of the Tresca yield criterion and of its associated flow rule. It is worth noticing that the utilisation of the above-mentioned yield criterion instead of more recent and sophisticated approaches based on non-quadratic yield functions (e.g. Wang and Pan, 2006) is due to its simplicity and ease of implementation for setting up an analytical model for the deformation mechanics of SPIF that is capable of dealing with both metals and polymers.

8.2.2 Mechanics of deformation

Previous research work by Franzen et al. (2008a, 2008b) and Silva et al. (2010) showed that formability in SPIF of polymers is limited by three different failure modes (Figure 8.2):

Figure 8.2 Failure modes that are experimentally observed in the SPIF of polymers: (a) Failure mode 1 (PC); (b) Failure mode 2 (PET); and (c) Failure mode 3 (PVC)

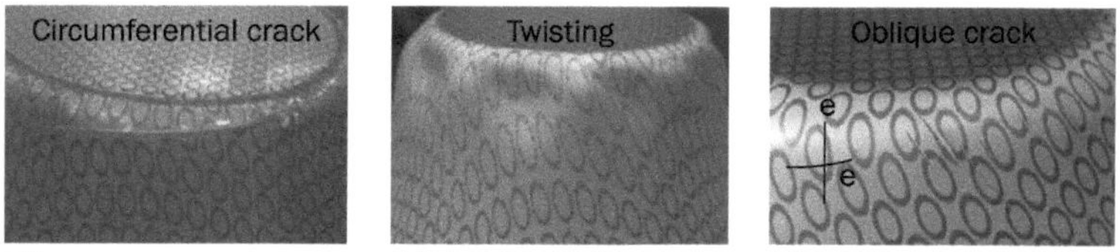

(a) (b) (c)

- Failure mode 1 – crack opening along the circumferential direction at the transition zone between the inclined wall and the corner radius of the parts (Figure 8.2a);
- Failure mode 2 – development of wrinkles along the inclined wall of the parts (Figure 8.2b);
- Failure mode 3 – crack opening along the bisector direction of the $\theta - \phi$ surface of the inclined wall of the parts (Figure 8.2c).

The mechanics of deformation associated with failure mode 1 is caused by stretching mechanisms due to meridional tensile stresses σ_{ϕ}. The morphology of the cracks and its propagation paths, along the circumferential direction, are similar to those usually found in the SPIF of metals.

Failure mode 2 is caused by wrinkles that are twisted about the axis of revolution in the direction of rotation of the forming tool and is triggered in the region of the inclined wall placed in the immediate vicinity of the corner radius, where thinning is more pronounced. The morphology of the wrinkles is a clear indication of difficulties in withstanding the twisting of radial cross-sectional planes arising from the movement of the forming tool along the circular path.

Failure mode 3 leads to a premature collapse of the SPIF parts and its mechanics of deformation seems to be related to the redundant straining that is produced as the forming tool progresses along the circumferential direction. Redundant straining is due to simultaneous straining by shearing on the $\theta - \phi$ surface and by bending, followed by additional unbending, along the θ direction. Silva et al. (2010) claimed that this failure mode is related to the initial surface integrity of the polymer sheets.

Strain and stress in the local region being plastically deformed

The theoretical framework for the SPIF of polymers is only able to model the deformation mechanics associated with failure mode 1, which is the more likely to develop during incremental sheet forming. The framework is built upon the extension of the membrane analysis with bi-directional in-plane contact friction forces, which was developed by Silva et al. (2008a) for the SPIF of metals, in order to include pressure-sensitive yield surfaces that are commonly found in the cold forming of polymers. Figure 8.3 shows the essentials for a membrane analysis in rotational symmetric conditions.

The forming tool acting on a polymer sheet is schematically illustrated in Figure 8.3a. The material is stretched by membrane forces, so that it conforms to the hemispherical shape of the tool, forming a contact area between the tool and the part of the sheet placed immediately ahead. The starting point of the analytical model to be presented is the assumption that the contact area between the single-point forming tool and the surface of the deforming sheet can be approximated by a small local shell element 'CDEF' with a shape resembling that of a radial slice (Figure 8.3b). The surface of revolution of the SPIF part is described by the rotation of the radial slice containing the shell element about the symmetry z-axis.

Figure 8.3 **Essentials for the theoretical framework based on membrane analysis: (a) Schematic representation of the local contact area between the tool and sheet placed immediately ahead; (b) Approximation of the local contact area by a shell element; (c) Cross-section view showing the acting stresses in meridional, circumferential and thickness directions**

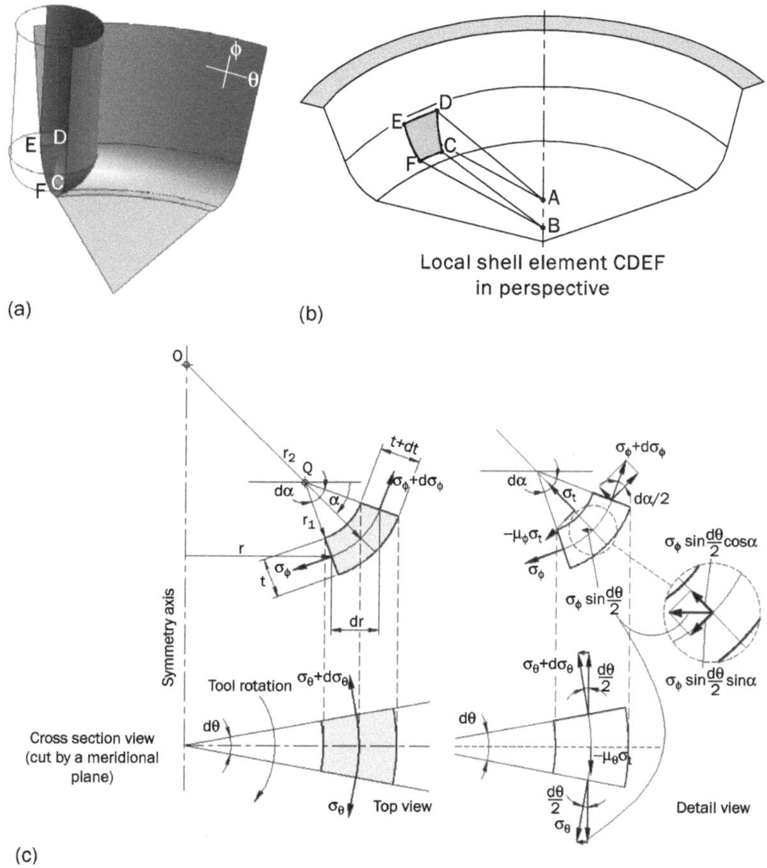

Figure 8.3c shows a cross-section view of the shell element obtained from a cut along the meridional plane. The normal to the surface of the shell element cuts the z-axis at the point O and r_2 is defined as the corresponding radius. The centre

of curvature lies at point Q on the normal and r_1 is the radius of curvature of the local shell element along the meridian direction. The shell element 'CDEF' is subjected to normal forces, shear forces and bending moments, so that it conforms to the hemispherical shape of the tip of the pin tool, forming a contact area between the tool and the part of the sheet placed immediately ahead of the moving tool. The state of strain and stress acting in the shell element can be derived from the membrane equilibrium conditions if bending moments are neglected and circumferential, meridional and thickness stresses are assumed to be principal stresses. Further simplifying assumptions are the following:

- The material is assumed as rigid-perfectly plastic, i.e. no strain-hardening effects are taken into consideration.
- The yield criteria and associated flow rules follow the modified versions of the Tresca and von Mises criteria that were proposed by Whitney and Andrews (1967), Raghava et al. (1973) and Caddell et al. (1974).
- The material is assumed isotropic, i.e. no anisotropic effects are taken into consideration.
- Rate effects are not taken into consideration because heat generation during SPIF is negligible and plastic deformation is expected to take place near or below the glass transition temperature.
- The resultant friction stress exerted at the tool–sheet contact interface is assumed to be made of two in-plane components – a meridional component – $\mu_\phi \sigma_t$ due to the step down movement of the tool and a circumferential component – $\mu_\theta \sigma_t$ due to circular movement of the tool combined with the rotational speed of the tool. This last assumption, which is an untraditional way of modelling

friction introduced for convenience, implies that the coefficient of friction $\mu = \sqrt{\mu_\phi^2 + \mu_\theta^2}$.

Resolving the force equilibrium along the thickness, circumferential and meridional directions results in:

$$\sigma_t\, r\, d\theta\, r_1\, d\alpha + \sigma_\phi\, r\, d\theta\, t \sin\frac{d\alpha}{2} + (\sigma_\phi + d\sigma_\phi)(r + dr)$$
$$d\theta(t + dt)\sin\frac{d\alpha}{2} + \sigma_\theta\, r_1\, d\alpha\, t \sin\frac{d\theta}{2}\cos\alpha + (\sigma_\theta + d\sigma_\theta) \quad (8.4)$$
$$r_1\, d\alpha\, t \sin\frac{d\theta}{2}\cos\alpha = 0$$

$$\sigma_\theta\, r_1 d\alpha\left(t + \frac{dt}{2}\right) - \mu_\theta \sigma_t\, r_1 d\alpha\left(r + \frac{dr}{2}\right) d\theta - (\sigma_\theta + d\sigma_\theta)$$
$$r_1 d\alpha\left(t + \frac{dt}{2}\right) = 0 \quad (8.5)$$

$$(\sigma_\phi + d\sigma_\phi)(r + dr)\, d\theta(t + dt) - \sigma_\phi\, r\, d\theta\, t + \mu_\phi\, \sigma_t\, r\, d\theta\, r_1\, d\alpha$$
$$-\sigma_\theta \frac{d\theta}{2} r_1\, d\alpha\, t \sin\alpha - (\sigma_\theta + d\sigma_\theta)\frac{d\theta}{2} r_1\, d\alpha\, t \sin\alpha = 0 \quad (8.6)$$

By disregarding higher order terms and simplifying the above equations one obtains:

$$\frac{\sigma_t}{t} + \frac{\sigma_\phi}{r_1} + \frac{\sigma_\theta}{r_2} = 0 \quad (8.7)$$

$$d\sigma_\theta = -\mu_\theta \sigma_t \frac{r d\theta}{t} \cong -\mu_\theta \sigma_t \quad (8.8)$$

$$\frac{d\sigma_\phi}{dr} + \frac{\sigma_\phi - \sigma_\theta}{r} + \frac{\mu_\phi \sigma_t}{t \sin\alpha} + \frac{\sigma_\phi}{t}\frac{dt}{dr} = 0 \quad (8.9)$$

Using the simplified form of the membrane equilibrium equation in the thickness direction (Equation 8.7) and taking into account the geometrical conditions prevailing in BC ($r_2 >> r_1$ and $r_2 >> t$; refer to Figure 8.3 for definition of these variables), it follows that the ratio between the circumferential stress σ_θ and the radius r_2 normal to the surface of the local shell element in Equation 8.7 can be disregarded. This leads to the following relation between the normal contact pressure $p = -\sigma_t$ acting on the local shell element, the meridional stress σ_ϕ, the thickness of the sheet t, and the radius of the forming tool ($r_1 = r_{tool}$):

$$\sigma_t = -\sigma_\phi \frac{t}{r_{tool}} \tag{8.10}$$

By considering the rigid-plastic constitutive equations derived from the associated flow rule (Equation 8.2) under plane strain conditions:

$$d\varepsilon_\theta = 0 \;\rightarrow\; \sigma_\theta = \frac{1}{2}\left(\sigma_t + \sigma_\phi - \left(\sigma_{YC} - \sigma_{YT}\right)\right) \tag{8.11}$$

and taking into account that typical ratios $R = \sigma_{YC}/\sigma_{YT}$ between the yield strengths of polymers in compression and tension are in the range of 1.1–1.4, it follows that the circumferential stress will take a value smaller than the arithmetic average of the other two membrane stress components (that is, $\sigma_\theta = \sigma_2 < 1/2\ (\sigma_t + \sigma_\phi)$). Thus, the principal stresses acting on the shell element 'CDEF' can always be identified as:

$$\begin{aligned} \sigma_\phi &= \sigma_1 > 0 \\ \sigma_\theta &= \sigma_2 \\ \sigma_t &= \sigma_3 < 0 \end{aligned} \tag{8.12}$$

where, $|\sigma_t| < |\sigma_\phi|$, because $r_{tool} > t$ (refer to Equation 8.10).

Although the relative values of the principal stresses acting on the shell element may vary with process parameters related to the thickness t of the sheet and to the radius r_{tool} of the single-point forming tool, the term $\bar{\sigma} = \sigma_1 - \sigma_3$ should always be kept equal to the flow stress of the polymer under perfectly plastic material assumptions. If the pressure-modified version of the Tresca yield criterion is to be considered, the flow stress can be defined as:

$$\bar{\sigma} = \sqrt{\sigma_{YC} \cdot \sigma_{YT} - (\sigma_{YC} - \sigma_{YT})\sigma_{kk}} = \sigma_Y \sqrt{1 - \beta\, \sigma_{kk}} \qquad (8.13)$$

where $\sigma_Y = \sqrt{\sigma_{YC} . \sigma_{YT}}$ is the geometric mean of the tensile and compressive yield strengths and $\beta = \left(\sigma_{YC} - \sigma_{YT}\right) / \sigma_Y^2$ is the normalised form of the strength-differential effect. Inserting Equation 8.13 in Equation 8.10 leads to the following equation for calculating the meridional stress σ_ϕ acting in the local shell element:

$$\sigma_\phi = \frac{\sigma_Y \sqrt{1 - \beta\, \sigma_{kk}}}{\left(1 + \dfrac{t}{r_{tool}}\right)} \qquad (8.14)$$

The thickness t is to be derived from $t = t_0 \sin \lambda$ according to the geometric relationship provided by the sine law (Silva et al., 2010).

Table 8.2 resumes the strains and stresses along the principal directions that were derived from the proposed analytical framework for the SPIF of polymers. It can be concluded that where the strength-differential effect is null, the distribution of strains and stresses for the SPIF of metals (Silva et al., 2008a) results.

Friction at the tool–sheet contact interface

Recent experimental results performed by Skjoedt et al. (2008a, 2008b) using single-point forming tools that are driven to rotate by the machine tool and that rotate free as a function of the contact with the surface of the sheet, reveal no influence of rotation in the overall formability of SPIF. Thus, it is possible to conclude that the influence of circumferential friction resulting from the contact between

Table 8.2 State of stress and strain in the small localised plastic zone of the rotational symmetric SPIF of metals and polymers

	State of strain	State of stress	Hydrostatic stress
SPIF – Metals (Silva et al., 2008a)	$d\varepsilon_\phi = -d\varepsilon_t > 0$ $d\varepsilon_\theta = 0$ $d\varepsilon_t < 0$	$\sigma_\phi = \sigma_1 = \sigma_Y / (1 + t/r_{tool}) > 0$ $\sigma_\theta = \sigma_2 = 1/2(\sigma_1 + \sigma_3)$ $\sigma_t = \sigma_3 = -\sigma_Y (t/r_{tool} + t) < 0$	$\sigma_m = (\sigma_Y/2)[(r_{tool} - t)/(r_{tool} + t)]$
SPIF – Polymers (Silva et al., 2010)	$d\varepsilon_\phi = -d\varepsilon_t > 0$ $d\varepsilon_\theta = 0$ $d\varepsilon_t < 0$	$\sigma_\phi = \sigma_1 = \left(\sigma_Y \sqrt{1 - \beta\sigma_{kk}}\right) / (1 + t/r_{tool}) > 0$ $\sigma_\theta = \sigma_2 = 1/2(\sigma_1 + \sigma_3 - \beta\sigma_Y)$ $\sigma_t = \sigma_3 = -\left(\sigma_Y \sqrt{1 - \beta\sigma_{kk}}\right)(t/r_{tool} + t) < 0$	$\sigma_m = \left(\sigma_Y \sqrt{1 - \beta\sigma_{kk}}/2\right)[(r_{tool} - t)/(r_{tool} + t)] - (1/6)\beta\sigma_Y$

the single-point forming tool and the sheet is negligible and only the friction forces exerted in the meridional direction need to be taken into account.

Under these circumstances, starting from the membrane equilibrium condition along the meridional direction, Equation 8.9, and disregarding the variation in thickness $dt/dr \cong 0$, it is possible to establish the following relationship for the distribution of the meridional stress σ_ϕ along the curvature from the bottom B to the top C of the local region being plastically deformed (Figure 8.4):

$$\sigma_\phi = \sigma_{\phi_B} \exp^{\mu\varphi} \qquad (8.15)$$

The symbol φ is defined in the detail of the SPIF part that is included in Figure 8.4. In physical terms, Equation 8.15 indicates that the meridional stress σ_ϕ will increase with r as a result of friction at the tool–sheet contact interface.

Figure 8.4 Schematic representation of the stress field in a cross-section view of the SPIF process by a meridional plane; the detail of the instantaneous plastically deforming region BC illustrates thinning at the corner radius

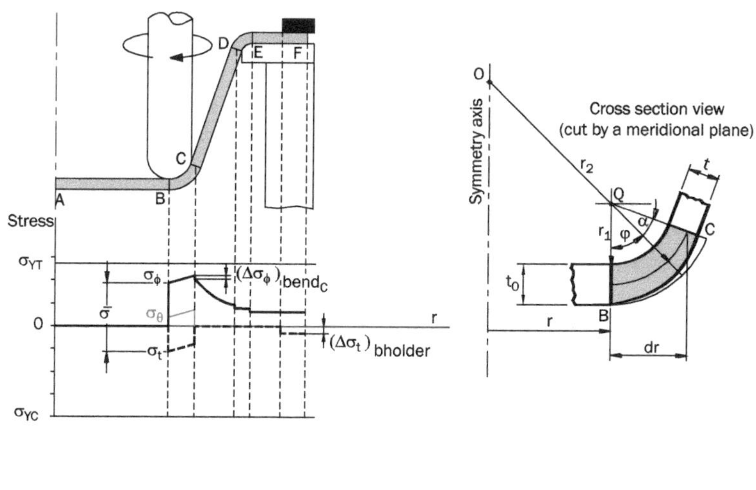

Thinning at the corner radius

Starting from the membrane equilibrium condition (Equation 8.9) in the meridional direction, disregarding friction and introducing the boundary conditions $r = r_B$, $\sigma_\phi = \sigma_{\phi B}$, and $t = t_0$, it is possible to derive the following identity:

$$\sigma_\phi = \sigma_{\phi_B} \frac{t_0}{t} \tag{8.16}$$

The physical meaning of Equation 8.16 is that reduction in thickness t tends to balance the rise in the meridional stress σ_ϕ so that $\sigma_\phi t$ is constant in the small localised region being plastically deformed. It follows that although the variation in the wall thickness of the inclined surface of the sheet can be approximately estimated by means of the sine law $t = t_0 \sin \lambda$, the reason why thinning occurs has to do with the meridional tensile stresses σ_ϕ rather than shearing acting in the small localised zone being plastically deformed. The detail included in Figure 8.4 provides a schematic, geometrical interpretation of thinning at the corner radius.

State of stress in the inclined wall adjacent to the forming tool

For the case of curvature $r_1 = \infty$ as well as the absence of contact pressure $\sigma_t = 0$ on the surface of the local element CD (see Figure 8.4), it may be concluded that $\sigma_\theta = 0$ along the inclined wall adjacent to the forming tool. Thus, since the meridional stresses σ_ϕ are the only stresses acting in this region, the following equilibrium condition is valid in the inclined wall of the sheet:

$$\sigma_{\phi_D} = \sigma_{\phi_C} \frac{r_C}{r_D} \tag{8.17}$$

According to Equation 8.17, the meridional stress σ_ϕ decreases along the inclined wall of the sheet, being higher at the transition point *C* and smaller at point *D*. Because the meridional stress at point *C* must be kept below the yield stress in tension (for a perfectly plastic material), it follows that the inclined wall surface of the sheet adjacent to the forming tool is elastic. This result, together with the stress field in the small localised plastic zone that is summarised in Table 8.2, results in the schematic distribution of the principal stresses that is depicted in Figure 8.4.

8.3 Experimental background

This section provides a brief summary of the procedures that were utilised in the independent determination of the most significant properties of the polymers and presents a brief overview of the experimental work plan on SPIF.

8.3.1 Material characterisation

The investigation was performed on commercial polymer sheet blanks having a uniform thickness profile of 2 and 3mm. Four different thermoplastics – polyethylene terephthalate (PET), polyamide (PA), polyvinylchloride (PVC) and polycarbonate (PC), with material structures varying from high-crystalline (PET) to amorphous (PC) – were employed.

The tensile yield stress σ_{YT} and the modulus of elasticity *E* of the polymers were determined by means of tensile tests. The specimens were machined from the supplied sheets in accordance with the ASTM D 638 norm and the testing speed was set to 5mm/min. The compression yield stress σ_{YC}

was determined by means of stack compression tests (Alves et al., 2011) that made use of multi-layer cylinder specimens that were assembled by piling up circular discs cut from the same sheet that was utilised for manufacturing the tensile test specimens.

Fracture toughness *R* was evaluated from tensile tests performed on double-edge notched (DEN) specimens by determining the specific essential work of fracture under limiting conditions of the length of the ligament being equal to zero (Martins et al., 2009).

The density of the polymers was measured from samples taken from the supplied sheets in a laboratory precision weighing scale equipped with a density determination kit. The experimental technique utilised in the determination of the densities was based on the Archimedes principle (or buoyancy method), which states that a body immersed in a fluid apparently loses weight by an amount equal to the weight of the fluid it displaces.

Table 8.3 presents the results of the independently determined physical and mechanical properties of the polymers.

8.3.2 Fracture forming limit diagram

Formability of the polymer sheet blanks was evaluated by means of tensile tests and bi-axial circular (50mm) and elliptical (50/32mm) hydraulic bulge tests. The tensile tests were performed in a mechanical testing machine while the hydraulic bulge tests were performed in a universal sheet metal testing machine. Testing speed was constant and set to 5mm/min in both cases.

Previous research work by the authors in PVC revealed that neck formation is suppressed in hydraulic bulge tests, so

Table 8.3 Representative summary of the physical and mechanical properties of the polymers utilised in the investigation with photographs of the specimens utilised in tensile and fracture toughness tests

Material	Structure	Elasticity modulus, *E* (MPa)	Yield stress at tensile, σ_{YT} (MPa)	Yield stress at compression, σ_{YC} (MPa)	Fracture toughness, *R* (kJ/m^2)	Density (kg/m^3)	*R*
PET	H-C	2,661.9	48.50	74.39	24.49	1.33	1.53
PA	M-C	927.9	29.50	32.00	26.46	1.14	1.09
PVC	L-C	3,491.9	34.50	58.56	24.28	1.43	1.70
PC	A	2,457.7	34.40	49.81	28.36	1.19	1.45

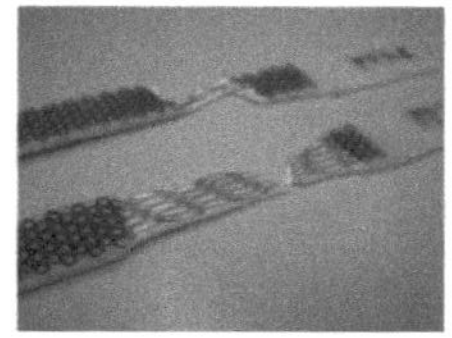

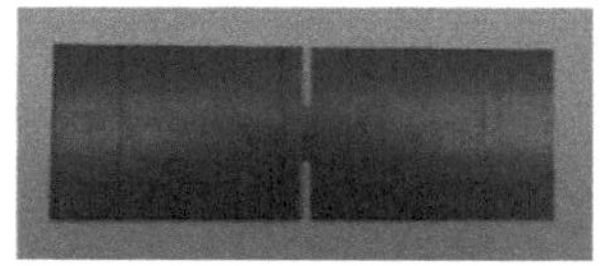

that traditional forming limit curves (FLCs) are inappropriate to describe failure and must be replaced by fracture forming lines (FFLs) (Silva et al., 2010). The technique utilised for obtaining the experimental FFL of a polymer requires measuring thickness before and after fracture in a microscope, at several places along the crack, in order to obtain the 'gauge length' strains (ε_1, ε_2). The strain in the width direction was obtained differently for tensile and bulge tests. In the case of tensile tests, measurements were directly taken from the width of the specimens whereas in the case of bulge tests measurements required the utilisation of the imprinted grid of circles in order to obtain the initial and deformed reference lengths. Details of the technique can be found elsewhere (Silva et al., 2009) and the FFLs of the polymers that were determined in the investigation can be approximated by straight lines falling from left to right, as shown in Figure 8.5.

8.3.3 Tool set-up and experimental work plan

The experiments were performed on a Deckel Maho CNC machining centre equipped with a simple tool set-up comprising the following components: (i) the blank holder with a backing plate and (ii) the single-point forming tool (Figure 8.1).

The blank holder is utilised for clamping and holding the polymer sheet in position during SPIF. The backing plate supports the sheet and its opening defines the working area of the single-point forming tool. The single-point forming tool has a hemispherical end and is utilised to progressively shape the polymer sheet into a component.

The experiments were performed in benchmark truncated conical and pyramidal shapes characterised by increasing

Figure 8.5 Fracture forming lines (FFL) of PC, PVC, PA and PET, inset shows pictures of tensile and bulge tests, and equations of approximation to the FFLs; schematic representation of the major and minor axis of the ellipses that result from the plastic deformation of the grids of circles

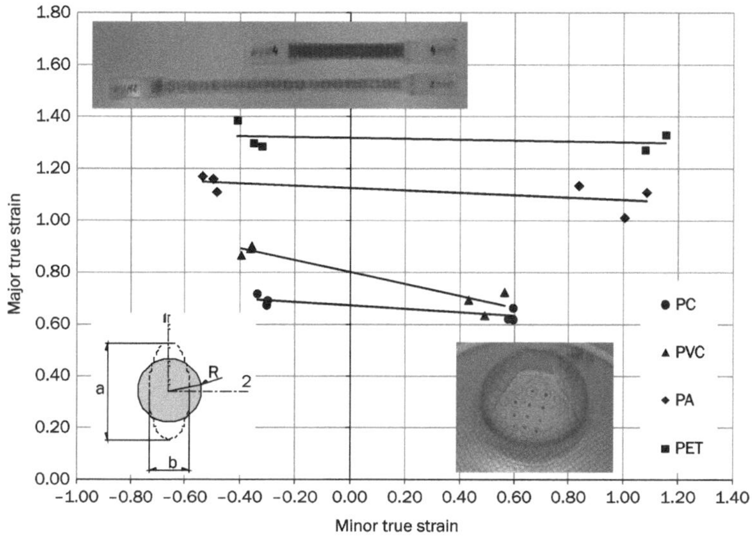

Material	Fracture forming line
PC	$\varepsilon_1 + 0.07\varepsilon_2 = 0.67$
PVC	$\varepsilon_1 + 0.30\varepsilon_2 = 0.80$
PA	$\varepsilon_1 + 0.04\varepsilon_2 = 1.13$
PET	$\varepsilon_1 + 0.02\varepsilon_2 = 1.32$

the drawing angle ψ with the depth (see Table 8.4) and the maximum drawing angle ψ_{max} was obtained from measuring the depth of the specimen at fracture. Test cases were selected in order to allow characterising the formability limits and to isolate the influence of the most relevant process parameters: (i) thickness of the sheet t_0, (ii) radius of the single-point forming tool r, and (iii) initial drawing angle ψ_0.

Table 8.4 Plan of experiments and geometrical details of the formability tests performed on truncated conical and pyramidal shapes with initial drawing angle ψ_0 and increasing drawing angle $\psi(h)$ with depth

Material	Initial drawing angle, ψ_0 (°)	Tool diameter (mm)	Geometry	Thickness of the sheet (mm)
PA, PC, PVC	30	8	Conical and Pyramidal	2
				3
		10	Conical	2
				3
		12	Conical and Pyramidal	2
				3
PET	30	8	Conical and Pyramidal	2
				3
		10	Conical	2
				3
		12	Conical and Pyramidal	2
				3
	45	8	Conical and Pyramidal	2
	60	8	Conical and Pyramidal	2
		10	Conical	2
		12	Conical	2

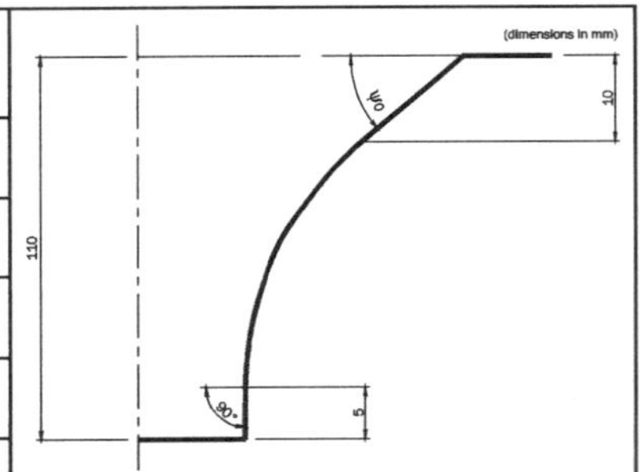

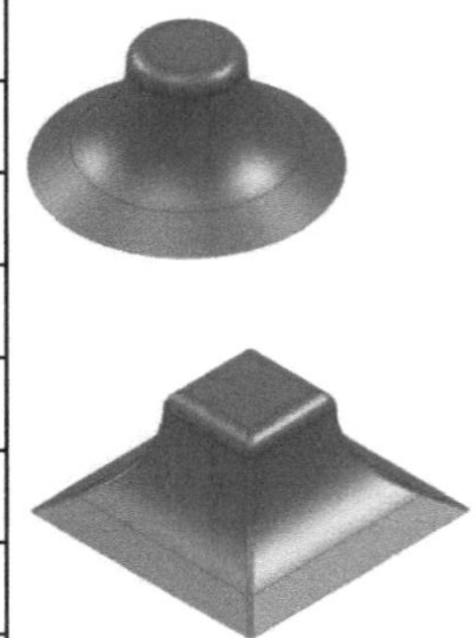

Three different single-point forming tools with hemispherical tips and radius of 4, 5 and 6mm were utilised. The tool path was helical with a step size per revolution equal to 0.1mm and was generated with the CAM module of the commercial software MasterCAM. The feed rate was set to 1,000mm/min and the lubricant applied between the forming tool and the polymer sheets was a water-soap emulsion in a ratio 1:1.

The technique utilised for obtaining the strains in the SPIF parts made use of the imprinted grid of circles with 1mm initial radius and involved measuring the major and minor axis of the ellipses that result from the plastic deformation of the circles during the tests. The values of strain were computed from (refer to Figure 8.5):

$$\varepsilon_1 = \ln\left(\frac{a}{2R}\right) \qquad \varepsilon_2 = \ln\left(\frac{b}{2R}\right) \tag{8.18}$$

where the symbol R represents the original radius of the circle and the symbols a and b denote the major and minor axis of the ellipse.

8.4 Results and discussion

The experimental distribution of the major and minor true strains obtained from circle grid analysis in the principal strain space is shown in Figure 8.6. The open markers refer to experimental strains that were taken along the meridional direction of the SPIF benchmark parts while the solid markers refer to experimental strains at failure. The latter were obtained from grid-elements placed just outside the crack since they represent the condition of the thinned sheet at the onset of fracture.

Figure 8.6 Experimental strains in truncated conical and pyramidal SPIF parts with 3mm thickness using an initial drawing angle $\psi_0 = 30°$: (a) PC, (b) PVC, (c) PA and (d) PET

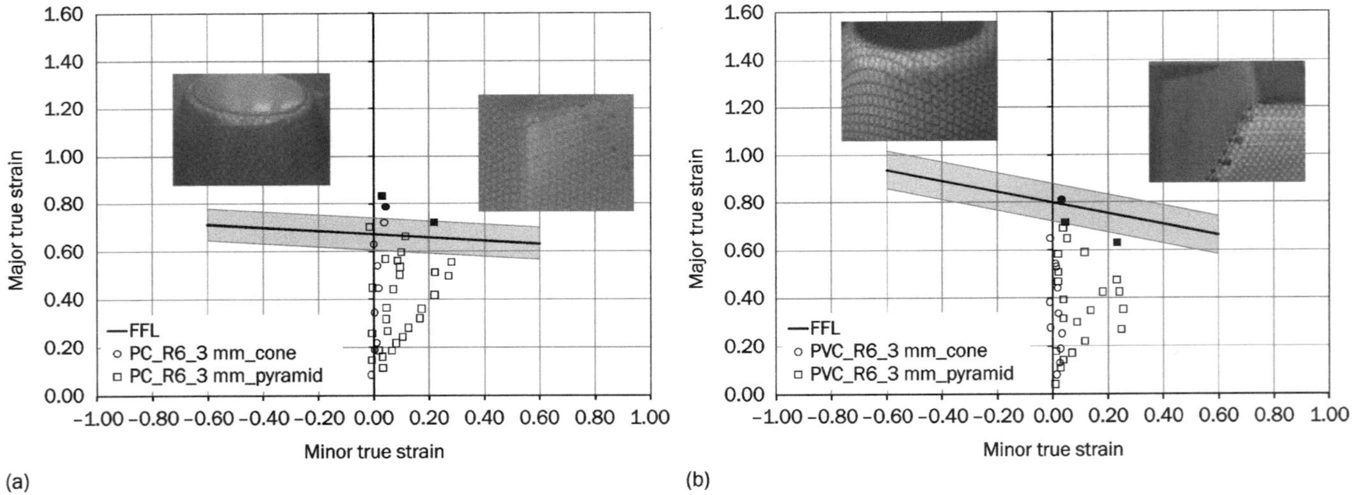

Figure 8.6 *Continued*

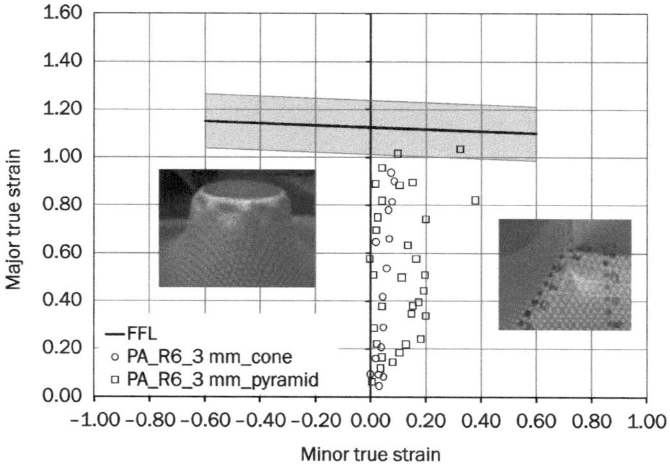

(c)

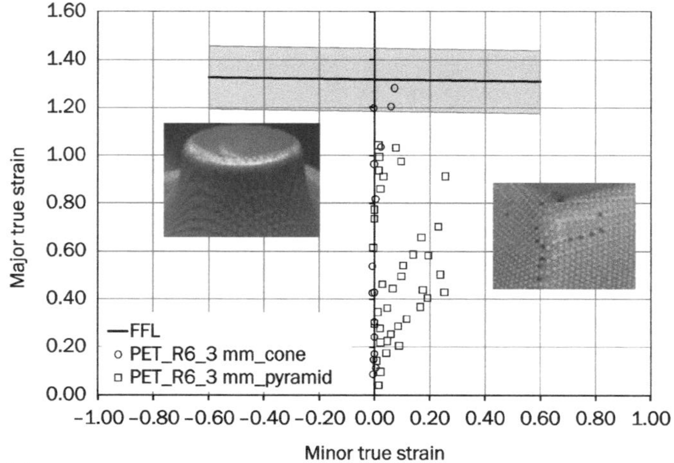

(d)

Results confirm that truncated conical shapes are formed under plane strain conditions while truncated pyramidal shapes are obtained under bi-axial stretching in the corners and plane-strain conditions in the side walls. It is worth noting that the experimental values of fracture strains for the pyramids are not placed on the equal bi-axial strain ratio line with slope +1 in the principal strain space. In fact, although the onset of failure is located at corners of the pyramids, the values of fracture strains are somewhat deviating towards the plane strain direction. This also explains the existence of different values of fracture strains plotted at various locations of the first quadrant of the principal strain space.

The black solid lines, denoted as 'FFL', are the fracture forming lines that were determined by means of tensile and bi-axial circular and elliptical hydraulic bulge tests (refer to Figure 8.5). The FFLs are bounded by grey areas corresponding to an interval of 10 per cent due to the experimental uncertainty in their determination.

The SPIF parts made from PC and PVC (Figures 8.6a and 8.6b) show a very good agreement between the experimental strains at failure and the FFL. In the case of the SPIF parts made from PA, the experimental strains at failure are higher than the corresponding FFL. This is because failure is not caused by crack opening along the circumferential direction (failure mode 1) but due to the development of wrinkles along the inclined wall of the parts (failure mode 2). The wrinkles are twisted about the axis of revolution in the direction of rotation of the single point forming tool and give rise to changes in the strain loading paths. For instance, the strain loading path of the conical SPIF part made from PA (Figure 8.6c) deviates from plane strain to bi-axial stretching conditions. The SPIF parts made from PET exhibit very high levels of formability. For instance, in the case of the parts formed with an initial drawing angle $\psi_0 = 30°$ no fracture was observed as the maximum drawing angle ψ_{max} measured in the experiments approaches 90°. This observation is in good agreement with the experimental values of strain being located below the FFL in the principal strain space (Figure 8.6d).

In contrast to metals where formability in SPIF is mainly influenced by the thickness of the sheet t, the radius of the forming tool r_{tool} and the ductility of the material, polymers may also show dependency on the initial value of the drawing angle, ψ_0 and to the strength-differential effect, β. All of these dependencies will be analysed in the following sections of the chapter.

8.4.1 Radius of the single-point forming tool

The experimental results presented in Figure 8.7 show the influence of tool radius r_{tool} on the formability of the process. As seen, the higher the radius of the tool the lower the formability of the SPIF parts made from PC and PVC. This conclusion corroborates the theoretical framework presented in 'Theoretical framework' because an increase in r_{tool} accounts for an increase in the triaxiality ratio σ_m/σ_Y (refer to Table 8.2 and Figure 8.8), which is directly related to the accumulation of ductile damage. In other words, the higher the radius r_{tool} the lower the formability and, therefore, the higher the experimental values of strain ε_1 (Figure 8.7).

The results obtained with PA and PET were omitted from the analysis because no fracture was observed during the experiments.

8.4.2 Thickness of the sheet

The increase of sheet thickness t has a positive influence on the overall formability of the SPIF of polymers,

Figure 8.7 Experimental strains obtained in the SPIF of truncated conical parts with 3mm thickness using an initial drawing angle $\psi_0 = 30°$ and tool radius r_{tool} = 4,5 and 6 mm: (a) PC, (b) PVC

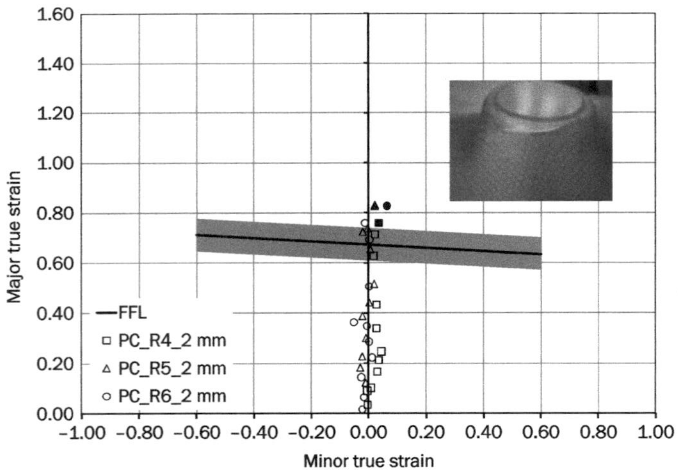

(a)

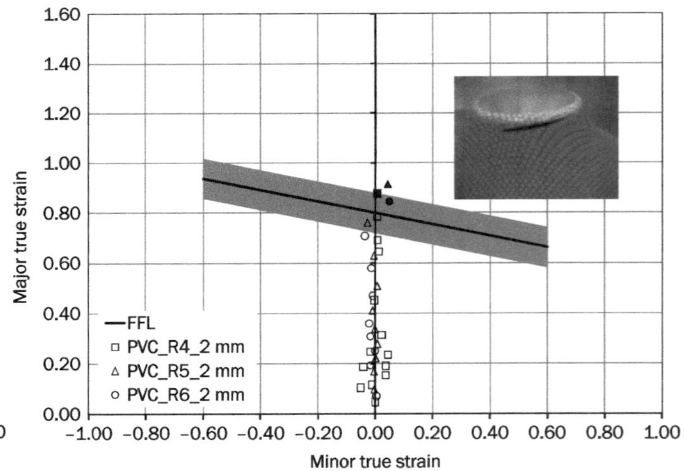

(b)

Figure 8.8 **Triaxiality ratio σ_m/σ_Y as a function of sheet thickness t calculated from the theoretical framework; the strength differential effect $\beta = 0$ for facilitating representation**

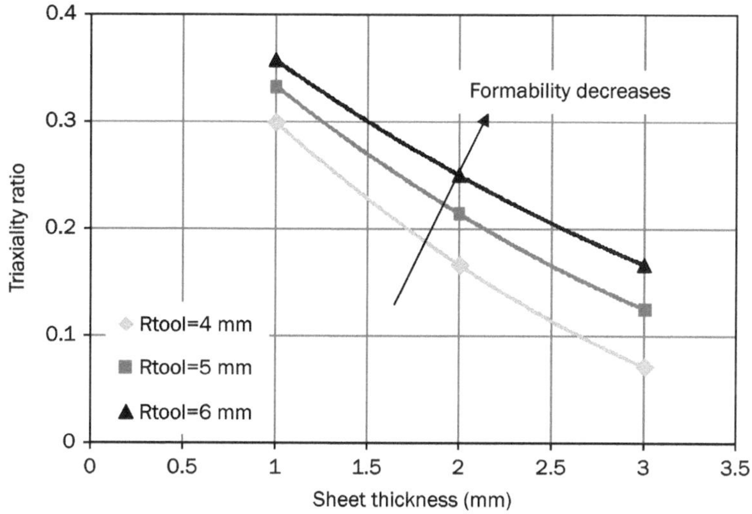

because it leads to small experimental values of strain ε_1 (Figure 8.9).

This result corroborates the estimate derived from the theoretical framework presented above, which accounts for a decrease in the triaxiality ratio σ_m/σ_Y as sheet thickness t increases (refer to Table 8.2 and Figure 8.8).

8.4.3 Drawing angle

In the SPIF of polymers there are two important drawing angles: (i) the initial drawing angle ψ_0 and (ii) the maximum drawing angle ψ_{max}. In relation to the initial drawing angle ψ_0, Figure 8.10 shows that in contrast to metallic materials, the formability of polymers is influenced by ψ_0 as it varies between 30° and 60°. On the contrary, the maximum drawing angle ψ_{max} of polymers is similar to those metallic materials, as can be seen in Table 8.5. In the case of PET, which presents

Figure 8.9 Experimental strains obtained in the SPIF of truncated conical parts with 2 and 3mm thickness using an initial drawing angle ψ_0 = 30° and a tool radius r_{tool} = 4mm. (a) PC, (b) PVC

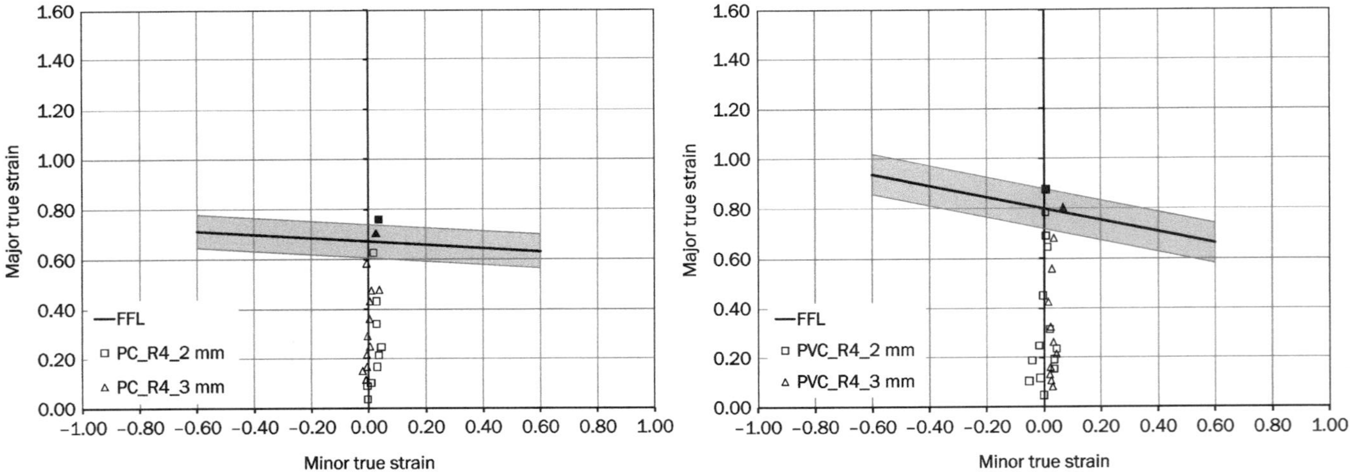

(a)

2 mm	3 mm
$\varepsilon_1 = 0.76$	$\varepsilon_1 = 0.70$

(b)

2 mm	3 mm
$\varepsilon_1 = 0.87$	$\varepsilon_1 = 0.80$

Figure 8.10 Experimental strains obtained in the SPIF of truncated conical parts with 2mm of thickness, made from PET, with a tool radius of 4mm and initial drawing angles ψ_0 of 30°, 45° and 60°

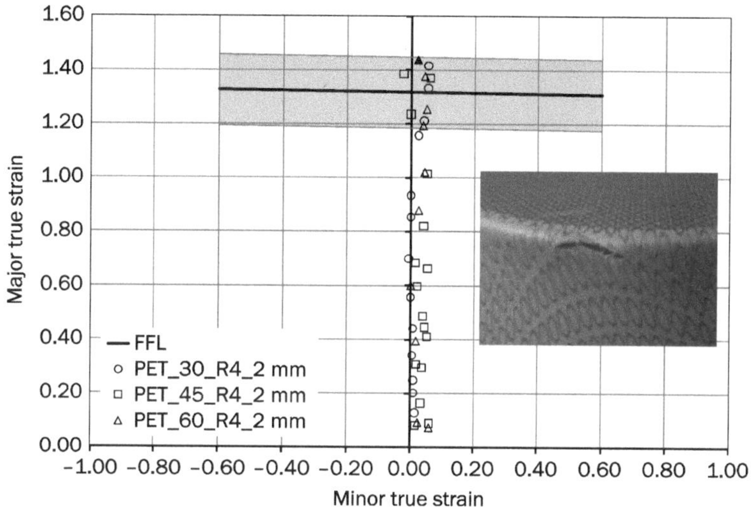

Table 8.5 Experimental maximum drawing angle ψ_{max} that sheet blanks made of steel, aluminum and polymers can undertake when forming a truncated conical shape by means of SPIF

Material		ψ_{max}
Steel	DC 01	76°
	DC 04	74°
High Strength Steel	H240	76°
	H400	72°
Aluminium	AA1050	80°
	AA5052	74°
Polymer	PET	~ 90°
	PC	72°
	PVC	66°

a maximum drawing angle very close to 90°, the outstanding formability may be justified by the strength-differential effect because an increase in β leads to a decrease in the hydrostatic stress σ_m and, therefore, to an increase in formability (refer to Table 8.2).

8.4.4 Variation in colour

SPIF parts made from PET, PA and PC experience no variation in density, allowing us to conclude that cold plastic deformation is incompressible (Figure 8.11). In relation to PVC, the variation in density and colour (from dark to light grey) is not due to plastic deformation but to the accumulation of crazes because, as recently mentioned by Bucknall (2007), crazing should be treated as a frustrated fracture process rather than a yield mechanism.

Figure 8.11 Variation in density as a function of the drawing angle ψ along the conical wall of SPIF parts made from PA, PC, PET and PVC (r_{tool} = 8*mm*, ψ_0 = 30°, t_0 = 2*mm*); inset shows a photograph of an SPIF part made from PVC

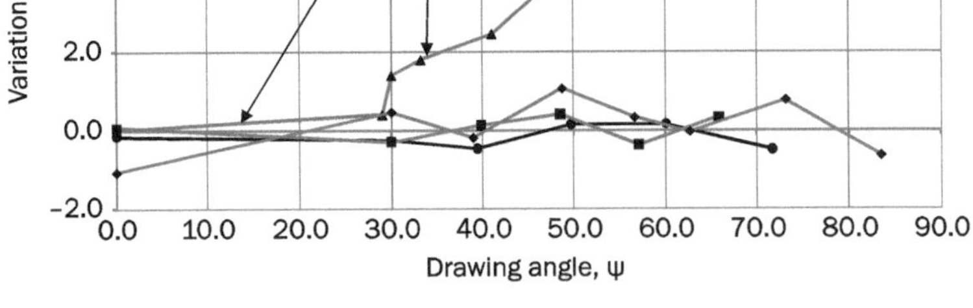

Crazing is known to take place when plastically deforming polymers below the glass transition temperature and is fundamentally related to the formation of a network of microvoids in the polymer matrix. As a result of this, the occurrence of crazing is accompanied by the occurrence of 'stress whitening' and by a significant and progressive reduction of density (Figure 8.11). As seen in Figure 8.11, the density of an SPIF conical part made of PVC is reduced up to a maximum value of 7 per cent near the crack.

The other polymers utilised in the investigation experienced almost no variation in colour and, in the case of PC, the sheet blanks remained transparent after being formed. This result opens a wide range of potential applications for the SPIF of industrial components made from PC.

8.5 Conclusions

This chapter evaluates the possibility of employing SPIF, currently utilised in metals, in the production of rapid prototypes and polymer sheet parts at room temperature, with large drawing angles and forming depths.

Results show that polymers add new phenomena in the characterisation of the formability limits of SPIF by introducing new modes of failure that are not experimentally observed in metals. The observations of crack opening along the circumferential direction at the transition zone between the inclined wall and the corner radius of the parts reveal that fracture is not preceded by localised necking and that crack propagates under tensile meridional stresses acting under stretching modes of deformation. These results are similar to those found by Silva et al. (2008a, 2008b) in the case of the SPIF of metals.

Experimental observations and measurements are interpreted by means of a theoretical framework that is capable of modelling the cold plastic deformation of polymers with pressure-sensitive yield surfaces. Qualitative evidence of the adequacy of the model to provide explanation of the results provides the link between theory and experimentation.

8.6 Acknowledgment

The authors would like to acknowledge PTDC/EME-TME/098728/2008 FCT/Portugal for the financial support of the research work.

8.7 References

Allwood, J.M., Shouler, D.R. and Tekkaya, A.E. (2007) The increased forming limits of incremental sheet forming processes, *Key Engineering Materials*, 344: 621–8.

Alves, L.M. and Martins, P.A.F. (2009) Cold expansion and reduction of thin-walled PVC tubes using a die, *Journal of Materials Processing Technology*, 209: 4229–36.

Alves, L.M., Nielsen, C.V. and Martins, P.A.F. (2011) Revisiting the fundamentals and capabilities of the Stack Compression Test, *Experimental Mechanics* (online first).

Bucknall, C.B. (2007) New criterion for craze initiation, *Polymer*, 48: 1030–41.

Caddell, R.M., Raghava, R.S. and Atkins, A.G. (1974) Pressure dependent yield criteria for polymers, *Materials Science and Engineering*, 13: 113–20.

Emmens, W.C. and van den Boogaard, A.H. (2007) Strain in shear and material behaviour in incremental forming, *Key Engineering Materials*, 344: 519–26.

Emmens, W.C. and van den Boogaard, A.H. (2009) An overview of stabilizing deformation mechanisms in incremental sheet forming, *Journal of Materials Processing Technology*, 209: 3688–95.

Eyckens, P., He, S., Van Bael, A., Van Houtte, P. and Duflou, J. (2007) Forming limit predictions for the serrated strain paths in single point incremental sheet forming, in *NUMIFORM '07; Proceedings of the 9th International Conference on Numerical Methods in Industrial Forming Processes*, Porto, Portugal.

Franzen, V., Kwiatkowski, L., Martins, P.A.F. and Tekkaya, A.E. (2008a) Single point incremental forming of PVC, *Journal of Materials Processing Technology*, 209: 462–9.

Franzen, V., Kwiatkowski, L., Neves, J., Martins, P.A.F. and Tekkaya, A.E. (2008b) On the capability of single point incremental forming for manufacturing polymer sheet parts, in *9th ICTP – International Conference on Technology of Plasticity*, Gyeongju, Korea.

Jackson, K. and Allwood, J.M. (2008) The mechanics of incremental sheet forming, *Journal of Materials Processing Technology*, 209(3): 1158–74.

Le, V.S., Ghiotti, A. and Lucchetta, G. (2008) Preliminary studies on single point incremental forming for thermoplastic materials, in *11th ESAFORM Conference on Material Forming – ESAFORM2008*, Lyon, France.

Lee, J.H. (1988) Upper bound analysis of the upsetting of pressure-sensitive polymeric rings, *International Journal of Mechanical Science*, 30: 601–12.

Martins, P.A.F., Kwiatkowski, L., Franzen, V., Tekkaya, A.E. and Kleiner, M. (2009) Single point incremental forming of polymers, *CIRP Annals – Manufacturing Technology*, 58: 229–32.

Raghava, R.S., Caddell, R.M. and Yeh, G.S.Y. (1973) The macroscopic yield behaviour of polymers, *Journal of Materials Science*, 8: 225–32.

Silva, M.B., Skjoedt, M., Martins, P.A.F. and Bay, N. (2008a) Revisiting the fundamentals of single point incremental forming by means of membrane analysis, *International Journal of Machine Tools & Manufacture*, 48: 73–83.

Silva, M.B., Skjoedt, M., Atkins, A.G., Bay, N. and Martins, P.A.F. (2008b) Single point incremental forming & formability/failure diagrams, *Journal of Strain Analysis for Engineering Design*, 43: 15–36.

Silva, M.B., Skjoedt, M., Bay, N. and Martins, P.A.F. (2009) Revisiting single-point incremental forming and formability/failure diagrams by means of finite elements and experimentation, *Journal of Strain Analysis for Engineering Design*, 44: 221–34.

Silva, M.B., Alves, L.M. and Martins, P.A.F. (2010) Single point incremental forming of PVC: Experimental findings and theoretical interpretation, *European Journal of Mechanics A/Solids*, 29: 557–66.

Skjoedt, M., Bay, N., Endelt, B. and Ingarao, G. (2008a) Multi stage strategies for single point incremental forming of a cup, *International Journal of Material Forming*, 1(1): 1199–202.

Skjoedt, M., Silva, M.B., Martins, P.A.F. and Bay, N. (2008b) Strain paths and fracture in multi stage single point incremental forming, in *9th ICTP – International Conference on Technology of Plasticity*, Gyeongju, Korea.

Spitzig, W.A. and Richmond, O. (1979) Effect of hydrostatic pressure on the deformation behaviour of polyethylene and polycarbonate in tension and in compression, *Polymer Engineering and Science*, 19: 1129–39.

Sternstein, S.S. and Ongchin, L. (1969) Yield criteria for plastic deformation of glassy high polymers in general stress fields, *Polymer Preprints – American Chemical Society*, 10: 1117–24.

Wang, D.A. and Pan, J. (2006) A non-quadratic yield function for polymeric foams, *International Journal of Plasticity*, 22: 434–58.

Whitney, W. and Andrews, R.D. (1967) Yielding of glassy polymers: volume effects, *Journal of Polymer Science – Part C*, 16: 2981–90.

Index

Kolibri 360 GmbH
Ferdinand-Jühlke-Straße 7
99095 Erfurt, Deutschland
produktsicherheit@kolibri360.de